# THE WAR IN THE AIR

Being the Story of
the part played in the Great War
by the Royal Air Force

APPENDICES
by
H. A. JONES

**The Naval & Military Press Ltd**

in association with

**The Imperial War Museum**
Department of Printed Books

Originally published 1937

Published jointly by
**The Naval & Military Press Ltd**
Unit 10 Ridgewood Industrial Park,
Uckfield, East Sussex,
TN22 5QE England
Tel: +44 (0) 1825 749494
Fax: +44 (0) 1825 765701

*and*

**The Imperial War Museum, London**
Depatrtment of Printed Books

# LIST OF APPENDICES

I. Memorandum on the organization of the Air Services, by Lieutenant-General Sir David Henderson, July 1917 . . . . . . 1

II. Air Organization. Second Report of the Prime Minister's Committee on Air Organization and Home Defence against Air Raids, dated 17th August 1917 (Smuts's Report) . . . . . 8

III. Sir Douglas Haig's views on a separate Air Service, 15th September 1917 . . . . 14

IV. 'Munitions Possibilities of 1918.' Extract from a paper dated 21st October 1917 by Mr. Winston Churchill, Minister of Munitions . . . . 18

V. The Bombing of Germany. Copy of a memorandum handed by Major-General H. M. Trenchard to the Prime Minister, Mr. Lloyd George, January 1918 . 22

VI. Memorandum on Bombing Operations. Forwarded by C.I.G.S., War Office, to General Sir Henry Wilson, British Military Representative, Supreme War Council, January 1918 . . . . 24

VII. Memorandum by Sir William Weir, Secretary of State for the Royal Air Force, on the responsibility and conduct of the Air Ministry, May 1918 . . 26

VIII. Memorandum by Marshal Foch on the subject of an Independent Air Force, 14th September 1918 . 29

IX. Joint Note No. 35. 'Bombing Air Force.' Addressed to the Supreme War Council by the Military Representatives . . . . . . 30

X. The Bombardment of the Interior of Germany. Memorandum by Marshal Foch, based on Joint Note No. 35 . . . . . . 30

XI. Heads of Agreement as to the constitution of the Inter-Allied Independent Air Force. An agreement reached between the British and French Governments and transmitted, through the Supreme War Council, to the American and Italian Governments for approval 41

XII. Statistics of work of squadrons of the Independent Force, including wastage. June–November 1918 . facing p. 41

XIII. Industrial Targets bombed by squadrons of the 41st Wing and the Independent Force. October 1917–November 1918 . . . . . 42

## LIST OF APPENDICES

XIV. Volklingen Steel Works. Analysis of Damage caused by Air Raids in 1916, 1917, and 1918 . . . 85

XV. State of Independent Force, Royal Air Force . . 87

XVI. Organization of Royal Air Force, Middle East, 30th September 1918 ⎫

XVII. Summary of Anti-Submarine Air Patrols from 1st May to 12th November 1918. (Home Waters) . . . . . ⎬ between pp. 87–88

XVIII. Comparison of Anti-Submarine Flying Operations, between Groups Nos. 9, 10, and 18. From 1st July to 30th September 1918 . . . . 88

XIX. A short review of the situation in the air on the Western front and a consideration of the part to be played by the American aviation. Memorandum by Headquarters, R.F.C., France, December 1917 . . 89

XX. Fighting in the Air: Memorandum issued by British G.H.Q., France, February 1918 . . . 92

XXI. Bombing Operations. Memorandum submitted to G.H.Q., by Major-General J. M. Salmond, G.O.C., Royal Air Force, France: June 1918 . . 110

XXII. Protection against Enemy Aeroplanes. Translation from a German document, July 1918 . . 113

XXIII. Methods of Bombing. Report from the Experimental Station, Orfordness: October 1918 . . . 114

XXIV. Order of Battle of the Royal Air Force, France, on 8th August 1918 . . . . . 116

XXV. The Battle of Amiens. Memorandum by G.O.C. V. Brigade, Royal Air Force. 14th August 1918 . 123

XXVI. Strength of the Royal Air Force, Western front, including Independent Force and 5th Group. 11th November 1918 . . . . . 125

XXVII. Types of Aircraft, 1914–18: Technical Data; Table A, Aeroplanes. Table B, Seaplanes and Ship Aeroplanes . . . facing p. 130

XXVIII. List of Squadrons, Royal Flying Corps and Royal Air Force, which served on the Western front, 1914–18 130

XXIX. List of Naval Squadrons which served with the Royal Flying Corps and Royal Air Force on the Western front, 1914–18 . . . . . 142

XXX. Location of R.A.F. units, Western front, 11th November 1918 . . . . . 144

# LIST OF APPENDICES

## STATISTICAL SECTION

| | | |
|---|---|---|
| XXXI. | British Aircraft produced and Labour employed, August 1914 to November 1918. (Figures for Germany, France, Italy, and America given, where available, for comparison) | 154 |
| XXXII. | Price list of various British war-time airframes and engines | 155 |
| XXXIII. | Firms and Labour employed on British aircraft production (excluding airships). Comparative detailed statement for the years 1916, 1917, and 1918 | 158 |
| XXXIV. | British naval airships built 1914–18 | 159 |
| XXXV. | Strength of British air personnel August 1914 and November 1918 | 160 |
| XXXVI. | Total casualties, all causes, to air service personnel, British and German, 1914–18 | 160 |
| XXXVII. | Comparison, by months, of British flying casualties (killed and missing) and hours flown on the Western front, July 1916 to July 1918) | 161 |
| XXXVIII. | Deliveries of anti-aircraft guns and ammunition (excluding naval) 1916 to 1918 | facing p. 161 |
| XXXIX. | Number of British anti-aircraft guns on the Western front, including Independent Force, July and November 1918 | 162 |
| XL. | Strength of Allied Aircraft on all fronts: June 1918 | between pp. 162–3 |
| XLI. | Disposition of aircraft and engines on the charge of the Royal Air Force at 31st October 1918. Table A, Aeroplanes and Seaplanes (Airframes). Table B, Engines | |
| XLII. | Length of front held by British in France, various dates, 1917 and 1918 | 162 |
| XLIII. | Hostile Bombing Activity on British front in France, May to October 1918 | 163 |
| XLIV. | Summary Statistics of German air raids on Great Britain, 1914–18 | 164 |
| XLV. | Anti-Aircraft Defences in Great Britain. Schedule of types, disposition, and strengths of aircraft, guns, height-finders, searchlights, and sound-locators, and strength of personnel, 10th June 1918 | 165 |
| XLVI. | Strength in personnel of Royal Air Force in various theatres of war at 31st October 1918 | 172 |

# APPENDIX I

## MEMORANDUM ON THE ORGANIZATION OF THE AIR SERVICES

*By Lieutenant-General Sir David Henderson, July 1917*

THE Royal Flying Corps came into existence in the month of May 1912. In its original organization it was intended to be a joint service, and was divided into a Naval and a Military Wing, the Central Flying School, the Royal Aircraft Factory, and a Reserve. The intention of the Sub-Committee of the Committee of Imperial Defence which drew up the original scheme, that the Corps should be a joint Corps, is evident from the following quotations from its Report, dated 28th February 1912:

'While it is admitted that the needs of the Navy and Army differ, and 'that each requires technical development peculiar to sea and land warfare 'respectively, the foundation of the requirements of each service is iden-'tical, viz. an adequate number of efficient flying men. Hence, though 'each service requires an establishment suitable to its own special needs, 'the aerial branch of one service should be regarded as a reserve to the 'aerial branch of the other. Thus in a purely naval war the whole of the 'Flying Corps should be available for the Navy, and in a purely land war 'the whole corps should be available for the Army. . . .'

'The British aeronautical service should be regarded as one, and should 'be designated "The Royal Flying Corps". The Flying Corps should 'supply the necessary personnel for a Naval and a Military Wing, to be 'maintained at the expense of, and to be administered by, the Admiralty 'and the War Office respectively. The corps should also provide the neces-'sary personnel for a Central Flying School, and a reserve on as large a 'scale as may be found possible.'

These different establishments, however, were separately controlled. The Military Wing, Central Flying School, the Royal Aircraft Factory, and the Reserve—with the exception of that portion of it which was composed of officers and men of the regular Naval Service—were under the administration of the War Office; the Naval Wing and the Naval Officers and men of the reserve were to be administered by the Admiralty. In order to ensure co-operation between the two services, a Joint Committee was formed called the Air Committee, composed as follows:

The Parliamentary Under-Secretary of State for War (Chairman).
The Commandant of the Central Flying School.
The Officer Commanding the Naval Wing of the Flying Corps.
The Commandant of the Military Wing of the Flying Corps.
The Director of the Operations Division, War Staff, Admiralty.
The Director of Military Training, General Staff, War Office.
The Director of Fortifications and Works, War Office.
The Superintendent of the Aircraft Factory.
Joint Secretaries { A Member of the Secretariat of the Committee of Imperial Defence.
An Officer of the Naval Flying Staff.

## THE ORGANIZATION OF

In practice this Committee proved somewhat unwieldy, and as might be expected from its composition, the members were inclined to range themselves into two parties, a Naval and a Military. The only member who, from his position, was entirely unprejudiced was the Commandant of the Central Flying School, who happened to be a Naval Officer—the present fifth Sea Lord of the Admiralty. The Meetings of the Committee, however, were of some value in acquainting each service with what the other was doing, but beyond this no very practical results were achieved from it.

In fact, the two services from the beginning tended to drift apart, rather than come more closely together. The Joint Service was never more than a pious aspiration, and when in 1914 the Naval Wing of the Royal Flying Corps was transformed into the Royal Naval Air Service, the separation became more marked. The Navy had always retained their old Flying School at Eastchurch, so that the one remaining bond of union, the Central Flying School, was not the only source from which the Naval Air Service drew its pilots. The two services were in this state of almost complete separation when the war broke out.

As soon as hostilities began, the necessity for rapid expansion of the Air Services, both for the Navy and Army, became apparent, and competition was inevitable. In the matter of personnel, the competition was not serious, although there was a great disadvantage in the fact that applicants who had been refused by one service were sometimes accepted by the other, but in the matter of supply, competition was really serious from the beginning, and both services suffered. An effort was made to eliminate this rivalry by dividing up the aeroplane and engine firms in the country between the Royal Flying Corps and the Royal Naval Air Service, but this was a rough-and-ready cut and not satisfactory. In the purchasing of aeroplanes and engines from France, there was direct competition, and there was no method discovered by which independent arbitration between the two services could be brought to bear. The Navy complained that there was really enough material available for both services, which then were very small, but that the methods of purchase by the Royal Flying Corps were slow and inefficient, and that therefore the Army were always trying to grab from the Navy material which the latter had been able to acquire. The Army, on the other hand, complained that the Navy purchased everything in sight, whether they required it or not, that their needs were not nearly so great as those of the Army, and that the material which they were purchasing was not required for proper Naval purposes. These views on both sides may have been justified or not, but there was a good deal of friction amongst subordinates, and not very much good feeling between the personnel of the services.

The necessity for some arbitration between the services, in the matter of supply, was brought to the notice of the War Council. The old Air Committee had from the beginning of the war been dormant, and although never abolished had never met. The new Air Committee under the Presidency of Lord Derby was instituted, but the same defects as with the old became apparent; the Committee had no real power, and was terminated

## THE AIR SERVICES

by the resignation of the Chairman. This effort had done no good, so a somewhat more definite attempt was made by the formation of the First Air Board, under the Presidency of Lord Curzon. This Board necessarily took some time to become acquainted with the situation, and to acquire sufficient technical knowledge to know exactly how matters stood. After some months of deliberation, for indeed the Board had no executive power, the Chairman issued a Report recommending a considerable extension of the powers of the Board, in order to enable it to deal with the situation. After considerable discussion, the War Cabinet appointed a second Air Board—the present one—with extended powers, but still so limited that it is only by the exercise of the utmost goodwill by the members of the Board that business can be properly carried on.

The present situation is this: The Board is responsible indirectly, through the Ministry of Munitions, for the supply of all aeroplanes and engines, and many of the accessories of aviation. It is permitted to discuss matters of policy, and to make recommendations thereon to the Board of Admiralty and the Army Council. The rest of the business of aeronautics is still divided between the Navy and the Army, and this remainder includes such important features as the provision of the whole of the personnel, the provision of aerodromes, buildings, and storage accommodation, all training and discipline, all plans of aerial defence, and the distribution of the aerial forces. The Air Board has cognizance of these matters only through the Naval and Military members of the Board. From this it arises that on the adoption of any policy, the responsibility of carrying it out is completely divided at present. For instance: a large increase of the Royal Flying Corps had been sanctioned, the Air Board have to supply the material and the Army Council the personnel; and the division of responsibility goes farther than this. In order to train the pilots, for which the Army Council are responsible, the Air Board must supply training aeroplanes, for which they are responsible, so that if there is a shortage of training aeroplanes or engines, there will be a shortage of pilots, and if there is a shortage in pilots, all the efforts of the Air Board in providing war aeroplanes and engines will be wasted. Similarly, for the training of pilots new aerodromes are required; the Army Council has to provide these. If the Department of Fortifications and Works should fail to supply the aerodromes, then the Military Aeronautics Directorate will fail to supply the pilots, and again all the efforts of the Air Board will be wasted owing to a failure which is quite outside their control or jurisdiction.

Another important factor in the future of the Air Service is the position of the Airship and Balloon Branch. When the R.F.C. was first formed, all the airships in possession of the Government belonged to the Military Wing. On 1st January 1914 the whole of these were, by a Cabinet order, turned over to the Navy, and since then the Army has had no airships. The R.F.C., however, still use kite balloons and ordinary spherical balloons in common with the R.N.A.S., and the supply is somewhat complicated.

The real disadvantage of the present system is, however, that no complete view of aerial policy by the Air Board is possible, for the airships are entirely under the Admiralty. It seems probable that the possible use of airships and of aeroplanes or seaplanes coincide at certain points, and if so, it is evident that no complete air policy can be carried out except by a body which has control of both branches.

The Air Board is at present composed of:

A President.
A Parliamentary Secretary.
A Representative of the Board of Admiralty.
A Representative of the Army Council, and Two Representatives of the Ministry of Munitions,

the two latter being business men dealing with the business affairs of Supply. One of the duties of this body is to allocate aeroplanes and engines to the Navy and to the Army. It is evident from the composition of the Board that this allocation must, in the end, be made by the President and the Parliamentary Secretary. In order to enable them to arrive at a correct decision they have nothing to go on except the arguments of the Naval and Military Representatives, and their own common sense and judgement. They have no advisory staff whatever; the Technical Department of the Air Board is concerned only with the technical details of the aircraft and their accessories. In the event of a serious disagreement as to the allocation of aircraft, the Naval and Military representatives have a right to appeal to the War Cabinet through the Admiralty and War Office respectively, and hitherto such a right of appeal has never been exercised. This, however, is rather a tribute to the judgement of the President and the Parliamentary Secretary, and the goodwill of the Naval and Military representatives, than a satisfactory evidence that the organization is properly fitted to carry out its duties. In order to enable the President to consider the larger questions of aerial policy, and to give him the means of forming a correct judgement as to the relative importance of the different methods of employing aircraft, the Air Board ought to be equipped with a staff on the lines of the General Staff at the War Office, or the War Staff at the Admiralty. Under present conditions, however, the formation of such a staff would not be an easy matter.

The only persons with sufficient training and knowledge to undertake such work are either Naval or Military officers, and to place such officers in a position in which their advice might be directly contrary to the policy of their Naval or Military superiors is not quite a workable proposition. Until there is a prospect of a real career in the Air Force, it will be found very difficult to form a staff whose opinions could be accepted as being entirely fearless and unprejudiced.

It is difficult to indicate any method of overcoming the present illogical situation of divided responsibility in aeronautics, except by the formation of a complete department and a complete united service dealing with all operations in the air, and with all the accessory services which that expression implies. A department would have to be formed on the general lines of the Admiralty and War Office, with a full staff, and with full responsibility for war in the air. Undoubtedly some portion of our air forces must be considered as accessory to the Navy and to the Army, and such con-

## THE AIR SERVICES

tingents would have to be allotted according to the importance of the sea and land operations in progress, but it does not seem necessary that such contingents should be composed of Naval or Military personnel; any suggestions of that kind would only prolong the situation of divided responsibility. Individuals of such contingents might be officers or men of either service, but for their period of service with the Air Forces they would be lent to the Air Ministry, and would be in every respect completely under the control of that Ministry. It would be difficult just now, in the middle of a great war, to train all our pilots in such a manner that they would be equally fitted for land or sea service, and therefore, in the main, the contingents would be composed of the personnel at present employed with the Army and with the Navy. The whole system of training, however, especially in its earlier stages, should be completely unified; and it is only when pilots and mechanics begin to specialize, and then not in all cases, that it would be necessary to earmark them for service by sea or by land. The unification of training ought to have an immediate effect both on efficiency and economy.

It is, of course, evident that until the immediate needs of the Navy and Army can be supplied, there will be no central Air Force available for independent operations. So far as the Army is concerned, however, there is a reason to hope that its immediate needs for fighting, for reconnaissance, and for artillery work, will be met in the early months of next year, and that even then a considerable force of bombing machines will also be available. If the Air Ministry were in existence now, it would be its duty to look ahead and consider the best means of employing this Service, that is to say, considering for instance whether it could be better employed under the direct command of the Commander-in-Chief in France, or only under his nominal command, if serving in France, but strategically directed by the General Staff of the Air Ministry. For the present Air Board to undertake such a study would be very difficult, and probably not very useful. There is no staff available to consider such questions except the staffs of the Naval and Military members of the Board, and these officers sit on the Board mainly as representatives of the Board of Admiralty and the Army Council. All investigations of the kind at present are purely Naval or Military, and it is not to be expected that the opinions expressed should be entirely free from the Naval or Military bias of these separate departments.

Yet it is by no means clear that, even on the present programme, we shall reach the limit of desirable expansion in the air forces. The problems of the future should be attacked now, otherwise we may waste force when we have it, or we may want additional force when we might have had it.

Although logically the desirability of a separate unified Air Force is almost beyond dispute, yet in its formation many administrative difficulties will have to be overcome, and this will be particularly difficult in time of war. In the first place, the formation of a complete staff for the Ministry is necessary; a branch for general staff duties, a branch for personnel, a branch for general and technical material and armament, a branch for works and buildings, and a complete financial establishment. Although these staffs need not necessarily be very large in numbers, it would be difficult to lessen the number of branches, and the finding of experienced personnel capable of taking the necessary responsibility would not be an easy matter at this period of the war. Further, as the Air Service would have to be a third Military Service, and separate from the Army and the Navy, it would be necessary to draw up and pass a Discipline Act on the lines of the Naval Discipline Act or the Army Act; to draw up and promulgate King's Regulations and prepare a Pay Warrant. It must be evident that these duties could not be undertaken by the officers who are at present responsible for the administration of the Air Service. A considerable amount of outside assistance would be required, and even with all possible facilities, this work would add very heavily to the burden of the senior officers and officials who are now connected with aeronautics.

If a scheme for a United air service were definitely adopted by the Government and announced, and measures taken to put it into effect, it is not anticipated that there would be any serious difficulty over the transference of the personnel from their old services to the new. But in this matter a great deal would depend on the definite announcement that the Service was to come into being, and that the scheme would be carried through. Otherwise, it is very likely that there would be a considerable amount of doubt and hesitation among the personnel, which would detract from the efficiency of the Force while the transfer was in progress.

To put it shortly, the formation of a new Air Service, even in war time, is not impossible, and although in certain respects it might cause temporary dislocation and reduction of efficiency, it may be that the final results would entirely outweigh this. The principal objection to carrying it out in war time is that the load of responsible officers is already so heavy that the addition of the considerable amount of somewhat complicated and original administrative work might be a serious distraction.

What had hitherto been considered the main obstacle to the formation of a separate Air Service is, however, a disadvantage that appears rather in peace time than during war. It is due to the consideration that the endurance of any person under the continued strain of flying is normally limited. Before the war it was considered that the active life of a flyer was probably about four years, under peace conditions, but under the stimulus of war, a number of pilots have already exceeded this limit, and the actual period of which a pilot will be able to continue active work is rather a matter of speculation. Undoubtedly, however, there is a limit and, therefore, in so far as the flying officer is concerned, not a very good prospect of a life career in the Air Service. As officers rise in rank above the rank of Flight Commander, their flying is not, in war at any rate, so continuous, but this is chiefly for the reason that in war the responsibility of a Squadron Commander is heavy and his work on the ground is pretty arduous. There is not, in fact, much time for him to fly. In peace, however, in order to preserve his position and his authority, it will be necessary for a Squadron Commander to fly regularly. There are a certain number of ground appointments which will be open to pilots who have given up flying,

## THE AIR SERVICES

but these pilots are not always the most suitable persons for such appointments. The fact has therefore to be faced that, as far as we know at present, there will be a continual flow from the Air Service into civil life, of young men who have been forced to give up flying and for whom no other appointments in the Air Service can be found. In the course of their service, such men would have acquired a considerable amount of technical skill, which might be made available in certain branches of civil life, but even so, it will not be easy to make the prospect of the young officer of the Air Service comparable with the prospects of regular officers of the Navy or Army.

In order to inculcate discipline, to ensure a good system of administration in units, and to infuse a good spirit into the Air Service, it would be most advisable in the early stages to continue to borrow a considerable number of officers, non-commissioned officers or petty officers, from the Army and Navy. The excellent spirit of efficiency of the Royal Flying Corps at the present moment is due almost entirely to the influence of the very carefully selected officers and non-commissioned officers who joined it from the regular Army.

The administrative difficulties of forming a united Air Service in time of war have been indicated, but in this special project of combining the Naval and Military Air Services at the present moment there are many minor and petty difficulties which will have to be considered, if the final decisions are to be accepted with unanimity and approval, and without creating such friction and jealousy as would interfere with the work that has to be carried out till the war is over. The Air Services are composed very largely of quite young men, most of them of a very lively temperament. Such matters as the adjustment of a relative rank in the two Services, or the uniform to be worn, of the titles of rank in the various grades, will be discussed in the Services with a great deal of freedom, and a good deal of prejudice. They would be discussed also, with more prejudice and with less knowledge, in Parliament and in the Press, and our experience in the past does not lead to the belief that either of those bodies will give much assistance in smoothing over difficulties. Personally, I think that if it is decided to form a United Air Service, the more important decisions—if the matter is handled with judgement—will be accepted without much objection, but that there will be the most violent controversies over the petty details.

To sum up the whole question, it seems only right that I should give a personal opinion. I believe that to ensure the efficiency of the Air Services in future, they ought to be combined, and that they should be under the control of a Ministry with full administrative and executive powers. The difficulties of carrying out this policy are purely war difficulties, and I believe that if this policy be carried out during the war, there will be a temporary loss of efficiency, which will be most marked in the administrative offices at present in control, but which must be considered even in the units actually engaged in operations. To minimize this loss of efficiency it would be advisable to draw up a complete scheme for the Air organization, and to take advice on the legal and administrative questions which

# 8 ORGANIZATION OF THE AIR SERVICES

have here been touched on, before announcing any change of policy, and after the announcement of the policy, it would be necessary to disregard entirely amateur advice and suggestions, and to leave the Air Board and the Services to work out their own salvation as best they may.

The time which will be occupied in preparing for the change will be considerable, and actually the balance of advantage or disadvantage in making the change depends on the estimate of the Government as to the duration of the war. If it be anticipated that the war will continue until next June, I think the change should be made for reasons of efficiency. But if the war should stop near the end of this year, we should lose rather than gain by attempting at present to alter the present system. The decision, therefore, appears to me to be a speculative one, but only in point of time, for I am convinced that eventually a united, independent Air Service is a necessity.

(Sgd.) DAVID HENDERSON.
Lieut.-General. D.G.M.A.

19. 7. 17.

## APPENDIX II

### AIR ORGANIZATION

*(Second Report of the Prime Minister's Committee on Air Organization and Home Defence against Air Raids, dated 17th August 1917)*

1. The War Cabinet at their 181st Meeting, held on 11th July 1917, decided—

'That the Prime Minister and General Smuts, in consultation with 'representatives of the Admiralty, General Staff and Field-Marshal Com-'manding-in-Chief, Home Forces, with such other experts as they may 'desire, should examine—

'(1) The defence arrangements for Home Defence against air raids.
'(2) The air organization generally and the direction of aerial operations.'

2. Our first report dealt with the defences of the London area against air raids. The recommendations in that report were approved by the War Cabinet and are now in process of being carried out. The Army Council have placed at Lord French's disposal the services of General Ashmore to work out schemes of air defence for this area. We proceed to deal in this report with the Second Term of Reference: the air organization generally and the direction of aerial operations. For the considerations which will appear in the course of this report we consider the early settlement of this matter of vital importance to the successful prosecution of the war. The three most important questions, which press for an early answer, are:

(1) Shall there be instituted a real Air Ministry responsible for all air organization and operations?
(2) Shall there be constituted a unified Air Service embracing both the present Royal Naval Air Service and Royal Flying Corps? And

# AIR ORGANIZATION

if this second question is answered in the affirmative, the third question arises—

(3) How shall the relations of the new Air Service to the Navy and the Army be determined so that the functions at present discharged for them by the Royal Naval Air Service and Royal Flying Corps, respectively, shall continue to be efficiently performed by the new Air Service?

3. The subject of general air organization has in the past formed the subject of acute controversies which are now, in consequence of the march of events, largely obsolete, and to which a brief reference is here made only in so far as they bear on some of the difficulties which we have to consider in this report. During the initial stages of air development, and while the role to be performed by an Air Service appeared likely to be merely ancillary to naval and military operations, claims were put forward and pressed with no small warmth, for separate Air Services in connexion with the two old-established War Services. These claims eventuated in the establishment of the Royal Naval Aircraft Service and Royal Flying Corps, organized and operating on separate lines in connexion with and under the aegis of the Navy and Army respectively, and provision for their necessary supplies and requirements was made separately by the Admiralty and War Office and to provide a safeguard against the competition, friction, and waste which were liable to arise, an Air Committee was instituted to preserve the peace and secure co-operation if possible. When war broke out this body ceased to exist, owing to the fact that its Chairman and members nearly all went to the front, but after a time it was replaced by the Joint War Air Committee. The career of this body was, however, cut short by an absence of all real power and authority and by political controversies which arose in consequence. It was followed by the present Air Board, which has a fairly well-defined status and has done admirable work, especially in settling type and patterns of engines and machines and in coordinating and controlling supplies to both the Royal Naval Air Service and Royal Flying Corps.

4. The utility of the Air Board is, however, severely limited by its constitution and powers. It is not really a Board, but merely a Conference. Its membership consists almost entirely of representatives of the War Office, Admiralty, and Ministry of Munitions, who consult with each other in respect of the claims of the Royal Naval Air Service and Royal Flying Corps for their supplies. It has no technical personnel of its own to advise it, and it is dependent on the officers which the departments just mentioned place at its disposal for the performance of its duties. These officers, especially the Director-General of Military Aeronautics, are also responsible for the training of the personnel of the Royal Flying Corps Service. Its scope is still further limited in that it has nothing to do either with the training of the personnel of the Royal Naval Air Service or with the supply of lighter-than-air craft, both of which the Admiralty has jealously retained as its special perquisites. Although it has a nominal authority to discuss questions of policy, it has no real power to do so, because it has not the independent technical personnel to advise it in that respect, and

any discussion of policy would simply ventilate the views of its military and naval members. Under the present constitution and powers of the Air Board, the real directors of war policy are the Army and Navy, and to the Air Board is really allotted the minor role of fulfilling their requirements according to their ideas of war policy. Essentially the Air Service is as subordinated to military and naval direction and conceptions of policy as the artillery is, and, as long as that state of affairs lasts, it is useless for the Air Board to embark on a policy of its own, which it could neither originate nor execute under present conditions.

5. The time is, however, rapidly approaching when that subordination of the Air Board and the Air Service could no longer be justified. Essentially the position of an Air service is quite different from that of the artillery arm, to pursue our comparison; artillery could never be used in war except as a weapon in military or naval or air operations. It is a weapon, an instrument ancillary to a service, but could not be an independent service itself. Air service on the contrary can be used as an independent means of war operations. Nobody that witnessed the attack on London on 11th July could have any doubt on that point. Unlike artillery an air fleet can conduct extensive operations far from, and independently of, both Army and Navy. As far as can at present be foreseen there is absolutely no limit to the scale of its future independent war use. And the day may not be far off when aerial operations with their devastation of enemy lands and destruction of industrial and populous centres on a vast scale may become the principal operations of war, to which the older forms of military and naval operations may become secondary and subordinate. The subjection of the Air Board and service could only be justified on the score of their infancy. But that is a disability which time can remove, and in this respect the march of events has been very rapid during the war. In our opinion there is no reason why the Air Board should any longer continue in its present form as practically no more than a conference room between the older services, and there is every reason why it should be raised to the status of an independent Ministry in control of its own war service.

6. The urgency for the change will appear from the following facts. Hitherto aircraft production has been insufficient to supply the demands of both Army and Navy, and the chief concern of the Air Board has been to satisfy the necessary requirements of those services. But that phase is rapidly passing. The programme of aircraft production which the War Cabinet has sanctioned for the following twelve months is far in excess of Navy and Army requirements. Next spring and summer the position will be that the Army and Navy will have all the Air Service required in connexion with their operations; and over and above that there will be a great surplus available for independent operations. Who is to look after and direct the activities of this available surplus? Neither the Army nor the Navy is specially competent to do so; and for that reason the creation of an Air Staff for planning and directing independent air operations will soon be pressing. More than that: the surplus of engines and machines now being built should have regard to the strategical purpose to which they are going to be put. And in settling in advance the types to be built

## AIR ORGANIZATION

the operations for which they are intended apart from naval or military use should be clearly kept in view. This means that the Air Board has already reached the stage where the settlement of future war policy in the air war has become necessary. Otherwise engines and machines useless for independent strategical operations may be built. The necessity for an Air Ministry and Air Staff has therefore become urgent.

7. The magnitude and significance of the transformation now in progress are not easily realized. It requires some imagination to realize that next summer, while our Western Front may still be moving forward at a snail's pace in Belgium and France, the air battle-front will be far behind on the Rhine, and that its continuous and intense pressure against the chief industrial centres of the enemy as well as on his lines of communication may form an important factor in bringing about peace. The enemy is no doubt making vast plans to deal with us in London if we do not succeed in beating him in the air and carrying the war into the heart of his country. The questions of machines, aerodromes, routes, and distances, as well as nature and scope of operations require careful thinking out in advance, and in proportion to our foresight and preparations will our success be in these new and far-reaching developments. Or take again the case of a subsidiary theatre; there is no reason why we may not gain such an overpowering air superiority in Palestine as to cut the enemy's precarious and limited railways communications, prevent the massing of superior numbers against our advance, and finally to wrest victory and peace from him. But careful staff work in advance is here in this *terra incognita* of the air even more essential than in ordinary military and naval operations which follow a routine consecrated by the experience of centuries of warfare on the old lines.

The progressive exhaustion of the man-power of the combatant nations will more and more determine the character of this war as one of arms and machinery rather than of men. And the side that commands industrial superiority and exploits its advantages in that regard to the utmost ought in the long run to win. Man-power in its war use will more and more tend to become subsidiary and auxiliary to the full development and use of mechanical power. The submarine has already shown what startling developments are possible in naval warfare. Aircraft is destined to work an even more far-reaching change in land warfare. But to secure the advantages of this new factor for our side we must not only make unlimited use of the mechanical genius and productive capacity of ourselves and our American allies, we must create the new directing organization, the new Ministry and Air Staff which could properly handle this new instrument of offence, and equip it with the best brains at our disposal for the purpose. The task of planning the new Air Service organization and thinking out and preparing for schemes of aerial operations next summer must tax our air experts to the utmost and no time should be lost in setting the new Ministry and Staff going. Unless this is done we shall not only lose the great advantages which the new form of warfare promises but we shall end in chaos and confusion, as neither the Army or Navy nor the Air Board in its present form could possibly cope with the vast developments

## AIR ORGANIZATION

involved in our new aircraft programme. Hitherto the creation of an Air Ministry and Air Service has been looked upon as an idea to be kept in view but not to be realized during this war. Events have, however, moved so rapidly, our prospective aircraft production will soon be so great, and the possibilities of aerial warfare have grown so far beyond all previous expectations, that the change will brook no further delay, and will have to be carried through as soon as all the necessary arrangements for the purpose can be made.

8. There remains the question of the new Air Service and the absorption of the Royal Naval Air Service and Royal Flying Corps into it. Should the Navy and the Army retain their own special Air Services in addition to the air forces which will be controlled by the Air Ministry? This will make the confusion hopeless and render the solution of the air problem impossible. The maintenance of three Air Services is out of the question, nor indeed does the War Office make any claims to a separate Air Service of its own. But, as regards air work, the Navy is exactly in the same position as the Army; the intimacy between aerial scouting or observation and naval operations is not greater than that between long-range artillery work on land and aerial observation or spotting. If a separate Air Service is not necessary in the one case, neither is it necessary in the other. And the proper and, indeed, only possible arrangement is to establish one unified Air Service, which will absorb both the existing services under arrangements which will fully safeguard the efficiency and secure the closest intimacy between the Army and the Navy and the portions of the Air Service allotted or seconded to them.

9. To secure efficiency and smooth working of the Air Service in connexion with naval and military operations, it is not only necessary that in the construction of aircraft and the training of the Air personnel the closest attention shall be given to the special requirement of the Navy and the Army. It is necessary also that all Air units detailed for naval or military work should be temporarily seconded to those services, and come directly under the orders of the naval or army commanders of the forces with which they are associated. The effect of that will be that in actual working practically no change will be made in the air work as it is conducted to-day, and no friction could arise between the Navy or Army commands and the Air Service allotted to them.

It is recognized, however, that for some years to come the Air Service will, for its efficiency, be largely dependent on the officers of the Navy and Army who are already employed in this work, or who may in the future elect to join it permanently or temporarily. The influence of the Regular officers of both services on the spirit, conduct, and discipline of the present air forces has been most valuable, and it is desirable that the Air Board should still be able to draw on the older services for the assistance of trained leaders and administrators. Further, it is equally necessary that a considerable number of officers of both Navy and Army should be attached for a part of their service to the Air Service in order that naval and military commanders and Staff Officers may be trained in the new arm and able to utilize to advantage the contingents of the air forces which will be put at

## AIR ORGANIZATION

their disposal. The organization of the air force therefore should be such as to allow of the seconding of officers of the Navy and Army for definite periods—not less than four or five years—to the Air Service. Such officers would naturally after their first training be chiefly employed with the naval and military contingents in order to secure close co-operation in air work with their own services. In similar fashion it would be desirable to arrange for the transfer of expert warrant and petty or non-commissioned officers from the Navy and Army to the new Service.

10. The summarize the above discussion we would make the following *recommendations:*

(1) That an Air Ministry be instituted as soon as possible, consisting of·a Minister with a consultative Board on the lines of the Army Council or Admiralty Board, on which the several departmental activities of the Ministry will be represented. This Ministry to control and administer all matters in connexion with aerial warfare of all kinds whatsoever, including lighter-than-air as well as heavier-than-air craft.

(2) That under the Air Ministry an Air Staff be instituted on the lines of the Imperial General Staff responsible for the working out of war plans, the direction of operations, the collection of intelligence, and the training of the air personnel; that this Staff be equipped with the best brains and practical experience available in our present Air Services, and that by periodical appointment to the Staff of officers with great practical experience from the front, due provision be made for the development of the Staff in response to the rapid advance of this new service.

(3) That the Air Ministry and Staff proceed to work out the arrangements necessary for the amalgamation of the Royal Naval Air Service and Royal Flying Corps and the legal constitution and discipline of the new Air Service, and to prepare the necessary draft legislation and regulations, which could be passed and brought into operation next autumn and winter.

(4) That the arrangements referred to shall make provision for the automatic passing of the Royal Naval Air Service and the Royal Flying Corps personnel to the new Air Service, *by consent*, with the option to those officers and other ranks who are merely seconded or lent, of reverting to their former positions.

There are legal questions involved in this transfer, and the rights of officers and men must be protected, but no dislocation need be anticipated.

(5) That the Air Service remain in intimate touch with the Army and Navy by the closest liaison, or by direct representation of both on the Air Staff, and that, if necessary, the arrangements for close co-operation between the three Services be revised from time to time.

(6) That the Air Staff shall, from time to time, attach to the Army and the Navy the air units necessary for naval or military operations, and such units shall, during the period of such attachment, be subject, for the purpose of operations, to the control of the respective naval

and military commands. Air Units not so attached to the Army and Navy shall operate under the immediate direction of the Air Staff.

The air units attached to the Navy and Army shall be provided with the types of machines which these services respectively desire.

(7) That provision be made for the seconding or loan of Regular officers of the Navy and Army to the Air Service for definite periods, such officers to be employed, as far as possible, with the naval and military contingents.

(8) That provision be made for the permanent transfer by desire, of officers and other ranks from the Navy and Army to the Air Services.

11. In conclusion, we would point out how undesirable it would be to give too much publicity to the magnitude of our air construction programme and the real significance of the changes in organization now proposed. It is important for the winning of the war that we should not only secure air predominance, but secure it on a very large scale; and having secured it in this war we should make every effort and sacrifice to maintain it for the future. Air supremacy may in the long run become as important a factor in the defence of the Empire as sea supremacy. From both these points of view it is necessary that not too much publicity be given to our plans and intentions which will only have the effect of spurring our opponents to corresponding efforts. The necessary measures should be defended on the grounds of their inherent and obvious reasonableness and utility, and the desirability of preventing conflict and securing harmony between naval and military requirements.

# APPENDIX III

### SIR DOUGLAS HAIG'S VIEWS ON A SEPARATE AIR SERVICE

15th September 1917.

The Chief of the
  Imperial General Staff.

IN accordance with your No. B.E.F./3/21, of the 24th August 1917, I submit the following remarks.

The principle of the formation of a separate Air Service having already been approved by the War Cabinet, my examination of the subject has been directed to the problem of ensuring that the efficiency of the Service now existing with our armies in the Field shall be maintained under the intended new organization. I have further limited my consideration of the matter to our requirements in this war, the winning of which demands the concentration of all our energies, while future needs can be foreseen more accurately and examined more closely after the war than is possible now.

As a first step towards offering suggestions as to how to apply the proposed change of organization to the Air Service with the armies in the

# SIR D. HAIG ON SEPARATE AIR SERVICE

field, without danger of causing loss of efficiency, I have carefully studied the report of the Committee, forwarded with your memorandum.

As a result of that study I may say at once that some of the views put forward as to future possibilities go far beyond anything that can be justified in my experience.

Apart from the question of advisability, from the point of view of morality and public opinion, of seeking to end the war by 'devastation of 'enemy lands and destruction of industrial and populous centres on a vast 'scale', I am unable to agree that there is practically no limit to such methods in this war, or that—at any rate in the near future—they are likely to 'become the principal operations of war, to which the older forms 'of military and naval operations may become secondary and subordinate'.

The first suggestion I have to offer, therefore, is that the limitations of long-distance bombing and the results to be expected from it may be very carefully examined in consultation with officers who have wide *practical knowledge* of its possibilities and of the general conditions affecting its development.

The scope of this examination should include a thorough consideration of the relative importance of independent bombing operations and of work in the air in combination with the older forms of military operations, as a means of winning this war.

I have no doubt that a full examination of these problems will show that the views expressed by the Committee require very considerable modification, and I desire to point out the grave danger of an Air Ministry, charged with such powers as the Committee recommends, assuming control with a belief in theories which are not in accordance with practical experience.

Long-distance bombing designed to cripple the enemy's naval and military resources and hamper his movements may certainly give valuable military results. The bombing of populous centres may also be justifiable, and may prove effective, in order to punish the enemy for similar acts previously committed by him, and to prevent their recurrence. Once such a contest is commenced, however, we must be prepared morally and materially to outdo the enemy if we are to hope to attain our ends.

Whatever room there may be for difference of opinion as to the results to be expected from long-distance bombing, there are very evident limitations to the possibilities of its execution in this war. The German air routes from Belgium to England are short and lie over the sea. The distances we have to go to reach important German centres are much greater and our routes lie over hostile territory. Our raiding machines, therefore, are far more likely to be seen and reported, and defence against them is more easily arranged for, both from the ground and in the air. In this regard it is well to remember that the science of defence against aircraft attack may develop considerably in the future.

Prevailing winds favour the enemy's return (the most dangerous part of the journey) from raids on England, whereas they will usually be against our machines returning from raids into Germany. This is a factor worthy of consideration as regards proportionate losses likely to occur.

The enemy's raids start from bases where he exercises absolute control.

Ours must start from bases situated in the territory of our allies which is already much congested by the troops of the various Allied Powers operating in France. It must be remembered in this connexion that, in addition to landing-grounds, every squadron in the air requires space for camps, workshops, &c., and the use of roads for the movement of transport.

Even in the air itself there is a limit to the number of machines that can be employed under war conditions, without confusion and regrettable mistakes, and experience shows that the chance of mistakes is so much increased when two or more independent services are working simultaneously that areas have to be carefully allotted to each.

In addition to all other considerations, the primary question of the supply of machines and trained personnel seems likely to cause limitations in the near future. After more than three years of war our armies are still very far short of their requirements, and my experience of repeated failure to fulfil promises as regards provision makes me somewhat sceptical as to the large surplus of machines and personnel on which the Committee counts in para. 3 of its report. Moreover that surplus is calculated on a statement of requirements rendered fifteen months ago, to which additions had to be made in November last and to which still further additions may have to be made. Nor is it clear that the large provision necessary to replace wastage has been sufficiently taken into account. The percentage of wastage has greatly increased this year as compared with last year and is likely to increase still further now that the enemy is straining every nerve to prevent our securing aerial ascendancy.

In regard to the general question of provision the views expressed by the Committee as to the relative importance of independent aerial operations and of the 'older forms' of military operations are of moment. The tendency of any authority holding such views must be to regard provision for these older forms of war as secondary and subordinate.

The amendment made to para. 10, sub-para. (6) of the Proceedings claims for the War Office and the Admiralty the right of judgement as to the requirements of the Army and Navy. It appears to me that this in effect must mean that there will in future be three authorities, each with its own special interests to satisfy; and, as pointed out in paras. 3 and 4 of the Committee's report, it has proved impossible up to date to adjudicate effectively between the claims of two, all efforts to create an authority capable of doing so having hitherto failed. In these circumstances there is room for doubt whether co-ordination under the new scheme will not be even more difficult than under the old, and whether the Air Ministry, claiming to be the supreme authority on aerial questions and to control the Air Services and holding the views quoted above, may not override military opinion as to military requirements.

It is so important that our armies in the field should be provided liberally with the means of gaining and maintaining supremacy in the air that the dangers I have outlined above of differences of opinion and conflict of interests cannot be too carefully guarded against.

A clear understanding is equally essential, firstly, as to the future relationship between a Commander of an army in the field and the Air Ser-

## SEPARATE AIR SERVICE

vices allotted to him; and, secondly, as to the relationship between him and any air forces working under an independent authority in or from the theatre in which he is operating.

As regards the first point, I observe that the Committee recommends that air units 'attached' to the Army for operations shall be subject to the military command for the purpose of operations, during the period of attachment. The Committee also recommends, however, that the Air Ministry shall 'control and administer all matters in connexion with aerial 'warfare of all kinds whatsoever', and that its air staff is to be 'responsible 'for the working out of war plans, the direction of operations, the collection 'of intelligence, and the training of the air personnel'. In face of these latter recommendations it is very essential, before the Air Ministry assumes control, to be quite clear as to what is meant by subjection to military command for the purpose of operations.

The Air Services with an army in the field are now as much a part of that army as are the infantry, artillery, or cavalry, and the co-ordination and combination of the efforts of all these services must be controlled directly by the Commanders of Armies, Corps, &c., under the supreme authority of the Commander-in-Chief.

To expect that the relationship between a Commander and the Army generally, on the one hand, and 'attached' units on the other, can ever be quite the same as if these units belonged to the Army, and looked to the other arms as their comrades and the Army authorities as their true masters and the ultimate judges on whom their prospects depend, would be contrary to all experience. No system of liaison or of seconding military officers to serve for a period of years with the Air Service can ever establish the same relationship as springs from community of interests and the knowledge that all belong alike to one service. It is therefore necessary to have a much fuller and clearer definition of the powers of military commanders in regard to air units placed at their disposal than is given in para. 10 of the Committee's report.

The military commanders must, of course, have disciplinary powers; but they should also, in my opinion, have power of reward and of selection, and at least a voice in the question of the transfer of units to and from their commands. Judgement as to the number of units required for each branch of air work—artillery observation, photography, fighting, and reconnaissance, &c.—as well as to the aerial methods to be employed in the field, must also be left to them if they are to be responsible for results.

It is also unsound to depart from the principle that the authority responsible for handling a service in the field should be charged with its training. The Committee proposes that the new air staff should be responsible for this, and, even apart from the principle involved, the requirements of the Army are now so varied, and go so much beyond mere flying, that it is a matter for consideration whether an Air Ministry could undertake training in all its branches. These already include gunnery, machine gunnery, observation of artillery fire, reconnaissance of troops, and other purely military subjects in addition to photography, wireless telegraphy, &c.

As regards the relationship between the Air Ministry and the

## 18 SIR D. HAIG ON SEPARATE AIR SERVICE

Commander in the field in respect of air units working under the former in or from an area controlled by the latter, the problem is also difficult and important. The allotment of objectives should be easy of adjustment. But the allotment of areas for aerodromes, &c., and of roads behind the allied lines in France, in a country already much congested, is a very different matter. In addition, the passage of aircraft controlled by an independent authority over the allied lines, through the areas where air activity is constant day and night and over anti-aircraft defences, and the avoidance of confusion and mistakes present difficulties to which at present I can see no satisfactory solution. The matter will require to be threshed out thoroughly by competent officers possessing practical knowledge of the difficulties to be overcome, and I must point out that during active operations such officers cannot easily be made available for the purpose from France.

Finally the question of kite balloons and the training of personnel for employment with them will require settlement.

I urge the importance of a thorough consideration of all these questions, and others which will doubtless arise, by a committee which shall include officers possessing considerable practical experience of the uses and conditions of employment of aircraft with an Army in the field. I also urge that this step may be taken before the responsibility for our air service is transferred, in order to guard against the danger of any temporary disorganization of the air services in the field during the transition period.

In conclusion I desire to add that I am in entire agreement with the view that the full development of all the possibilities of aerial attack is of urgent importance and that we must expect to have to meet considerable progress by the enemy in the same direction in the immediate future. It is essential, however, that our development should be on sound lines and based on practical experience, and it is with that object that I urge the need for a thorough investigation of the questions I have touched on above.

D. HAIG.
*Field Marshal, Commanding-in-Chief,*
*British Armies in France.*

## APPENDIX IV

### MUNITIONS POSSIBILITIES OF 1918

*Extract from a paper by Mr. Winston S. Churchill, Minister of Munitions, dated 21st October 1917*

*Section IV.* Most important of all the mechanical factors which are available comes the Air Offensive. So much progress in thought has been made on this subject, even since this paper was under preparation, that it is not necessary to dwell upon it at any length. But there are certain general principles which may be stated or restated.

War proceeds by slaughter and manœuvre. Manœuvre consists either in

## MUNITIONS POSSIBILITIES OF 1918

operations of Surprise or in operations against the flanks and communications of the enemy. Owing to the lines now stretching continuously from the Alps to the sea, there are no flanks. But the Germans, striking under the sea at our vital communications, have threatened us with a decisive peril, which we are warding off only by an immense diversion of our resources. If we take, on the one hand, the amount of national life-energy which the Germans have put into their submarine attack and compare it with the amount of national life-energy we are compelled to devote to meeting and overcoming that attack, it will be apparent what a fearfully profitable operation this attack on our communications has been to the enemy. Would it be an exaggeration to say that for one war-power unit Germany has applied to the submarine attack we have been forced to assign fifteen or twenty?

Even better than an operation against communications is an operation against bases. Air predominance affords the possibility of striking at both. It can either paralyse the enemy's military action or compel him to devote to the defence of his bases and communications a share of his straitened resources far greater than what we need in the attack.

All attacks on communications or bases should have their relation to the main battle. It is not reasonable to speak of an air offensive as if it were going to finish the war by itself. It is improbable that any terrorization of the civil population which could be achieved by air attack would compel the Government of a great nation to surrender. Familiarity with bombardment, a good system of dug-outs or shelters, a strong control by police and military authorities, should be sufficient to preserve the national fighting power unimpaired. In our own case we have seen the combative spirit of the people roused, and not quelled, by the German air raids. Nothing that we have learned of the capacity of the German population to endure suffering justifies us in assuming that they could be cowed into submission by such methods, or, indeed that they would not be rendered more desperately resolved by them Therefore our air offensive should consistently be directed at striking at the bases and communications upon whose structure the fighting power of his armies and his fleets of the sea and of the air depends. Any injury which comes to the civil population from this process of attack must be regarded as incidental and inevitable.

The supreme and direct object of an air offensive is to deprive the German armies on the Western Front of their capacity for resistance. It must therefore be applied and reach its maximum development in proper relation to the main battles both of Exhaustion and Surprise during the culminating period of our general offensive. German armies whose communications were continually impeded and interrupted and whose bases were unceasingly harried might still, in spite of all that could be done from the air, be able to maintain themselves in the field and keep the front. But if at the same time that this great difficulty and menace to their services in rear had reached its maximum, they were also subjected to the intense strain of a great offensive on the ground proceeding by battles both of Exhaustion and Surprise, the complete defeat and breaking up of their armies in the West as a whole might not perhaps be beyond the bounds of possibility. There is an immense difference between merely keeping an army fed and supplied on a comparatively quiescent front in spite of air attacks, and resisting the kind of offensive which the British are delivering at the present time. It is imperative that the defending army should be able to move hundreds of thousands of tons of stores and ammunition within very limited times to the battle front and to maintain a most rapid circulation of hundreds of thousands of troops; and the double strain of doing this under a really overwhelming air attack might well prove fatal. More especially might this be hoped for if the form of our offensive were not confined simply to the main battle front, but if it were so varied in locality and direction as to require from the enemy *an exceptional degree of lateral mobility.* For our air offensive to attain its full effect it is necessary that our ground offensive should be of a character to throw the greatest possible strain upon the enemy's communications.

We have greatly suffered and are still suffering in the progress of our means of air warfare from the absence of a proper General Staff studying the possibilities of air warfare, not merely as an ancillary service to the special operations of the Army or the Navy, but also as an independent arm co-operating in the general plan. Material developments must necessarily be misguided so long as they do not relate to a definite War Plan for the Air, which again is combined with the general War Plan.

In consequence of this many very important points are still in doubt or in dispute, on which systematized Staff study could have by now given clear pronouncements. The dominating and immediate interests of the Army and the Navy have overlaid air warfare and prevented many promising lines of investigation from being pursued with the necessary science and authority. Extreme diversities of opinion prevail as to the degree of effectiveness which can be expected from aerial attack. It is disputed whether air attack can ever really shatter communications, bases, or aerodromes. It is contended that aerodromes are difficult to discover and still more difficult to hit; that tons of bombs have been discharged on particular aerodromes without denying their use to the enemy; that railway junctions and communications have been repeatedly bombed without preventing appreciably the immense and continuous movement of men and material necessary to the fighting armies; that no bombardment from the air, especially at great distances from our own lines, can compare in intensity with the kind of bombardment from artillery, in spite of which, nevertheless, operations of a military and even semi-military character are continuously carried on.

On the other hand, it is claimed that aerial warfare has never yet been practised except in miniature; that bombing in particular has never been studied as a science; that the hitting of objectives from great heights by day or night is worthy of as intense a volume of scientific study as, for instance, is brought to bear upon perfecting the gunnery of the Fleet; that much of the unfavourable data accumulated showing the comparative ineffectiveness of bombing consists of results of unscientific action—for instance, dropping bombs singly without proper sighting apparatus or specially trained 'bomb droppers' (the equivalent of 'gun layers'), instead of dropping them in regulated salvos by specially trained men, so as to 'straddle' the targets properly. It is believed by the sanguine school that a very high degree of accuracy, similar to that which has been attained at sea under extraordinarily difficult circumstances, could be achieved if something like the same scientific knowledge and intense determination were brought to bear.

Secondly, it is pointed out that an air offensive has never been considered on the same scale or with the same ruthlessnes in regard to losses for adequate objects as prevail in the operations of armies. Aeroplanes have never been used to attack vital objectives in the same spirit as infantry have been used, viz. regardless of loss, the attack being repeated again and again until the objective is secured. It is pointed out that in 1918 numbers will for the first time become available for operations, not merely on the larger scale, but of a totally different character.

On the assumption that these more sanguine views are justly founded, the primary objective of our air forces become plainly apparent, viz. the air bases of the enemy and the consequent destruction of his air fighting forces. All other objectives, however tempting, however necessary it may be to make provision for attacking some of them, must be regarded as subordinate to this primary purpose. If, for instance, our numerical superiority in the air were sufficient at a certain period next year to enable us in the space of two or three weeks to sacrifice 2,000 or 3,000 aeroplanes, their pilots being either killed or captured, in locating and destroying by bomb and fire, either from a great height or if necessary from quite low down, all or nearly all the enemy's hangars, and making unusable all or nearly all his landing-grounds and starting-grounds within 50 or 60 miles of his front line, his air forces might be definitely beaten, and once beaten could be kept beaten.

Once this result was achieved and real mastery of the air obtained, all sorts of enterprises which are now not possible would become easy. All kinds of aeroplanes which it is not now possible to use on the fighting fronts could come into play. Considerable parties of soldiers could be conveyed by air to the neighbourhood of bridges or other important points, and, having overwhelmed the local guard, could *from the ground* effect a regular and permanent demolition. The destruction of particular important factories would also be achieved by carefully organized expeditions of this kind. 'Flying columns' (literally) of this character could be organized to operate far and wide in the enemy's territory, thus forcing him to disperse in an indefinite defensive good troops urgently needed at the front. All his camps, depots, &c., could be made the object of constant organized machine-gun attack from low-flying squadrons. But the indispensable preliminary to all results in the air, as in every other sphere of war, is to defeat the armed forces of the enemy.

# APPENDIX V

## THE BOMBING OF GERMANY

*Copy of a memorandum handed by Major-General H. M. Trenchard to the Prime Minister, Mr. Lloyd George, January 1918*

THERE appears to be in some quarters very serious misapprehension as to the extent to which Germany can be bombed during the spring and summer of this year. The American contribution will not have begun to mature; I am doubtful whether there will be a single American squadron actually engaged in bombing Germany by July. Our own efforts will be the greatest possible. The following paragraphs summarize what we can do.

1. *Aerodrome accommodation.*

Provision is being made in the Nancy neighbourhood for accommodating twenty-five squadrons by the beginning of May. In the meantime bombing operations are being carried on from Ochey, where ground is being temporarily lent to us by the French. It is hoped to have accommodation for fifteen more squadrons ready by the beginning of July, and still further accommodation in the same area soon afterwards. The difficulties of construction are very great, but, assuming that the labour necessary will continue to be forthcoming, the accommodation will more than keep pace with the provision of British bombing squadrons available to use it.

Aircraft Parks, a Repair Depot, and a Stores Depot for this southern area are also being hurried forward.

2. *Machines at present available.*

Germany is at present being bombed from the Ochey area by *one* Handley Page squadron (which is now short of three machines), *one* F.E.2b squadron (short-distance night flying only), and by *one* D.H.4 (R.R.) squadron.

It is worth noting that the bombing of Germany which has taken place in 1917, involving the dropping of about 20 tons of bombs in a few weeks in winter, has been done by this limited force, for though the figures of additional bombers about to be set out are meagre and disappointing, they represent some power of doing substantial mischief.

3. The following table shows what additional squadrons—Handley Pages and de Hav. 9's—will be available hereafter. The H.P. is the only type at present available to go long distances, or even as far as Mannheim, unless the weather is very favourable. The D.H.9 is as yet only in prospect. It is not anticipated that Germany will be bombed by more than the following squadrons at the dates given:

# THE BOMBING OF GERMANY

*Additional Squadrons for Bombing Germany*

The following squadrons will be actually at work as follows:

| | | |
|---|---|---|
| By 30th April . . . . | 1 H.P.* | 1 D.H.9† |
| By 31st May . . . . | 1 H.P. | 3 D.H.9 |
| By 30th June . . . . | 1 H.P. | 8 D.H.9 |
| By 31st July . . . . | 2 H.P. | 12 D.H.9 |
| By 31st August . . . | 5 H.P. | 17 D.H.9 |
| By 30th Sept. . . . . | 8 H.P. | 19 D.H.9 |
| By 31st Oct. . . . . | 11 H.P. | 19 D.H.9 |

\* This is the squadron already there, which has to be made up to the full strength.
† There is, moreover, one D.H.4 squadron already operating.

Though every effort will be made, and must be made to maintain this programme, its fulfilment necessarily depends upon the punctual delivery of machines and engines now promised by the Munitions Department. And on the assumption that these promises are completely performed, it must be accepted as a fact that it is impossible to create more squadrons for long-distance bombing than those above indicated before the dates given. Before a squadron can actually be used to bomb at a distance with a new type of machine much time is inevitably occupied in getting the machine to work satisfactorily and in the specially difficult task of training pilots. The 'teething' troubles which are always experienced with new machines and engines have to be completely eradicated if long journeys into enemy territory are to be safely and efficiently carried out.

This memorandum does not attempt to discuss why the figures are not larger, but the figures themselves cannot be successfully challenged. It is far better to know what can really be done so as to be able to count on it than to indulge in more generous estimates which cannot be realized.

*4. American co-operation.*

It is obvious by comparing the table in paragraph 3 with paragraph 1 that the aerodrome accommodation in the Nancy area will not be fully occupied by our machines for a long time. Accordingly, I discussed with the Americans in France arrangements to allow them to place one or more of their squadrons in the Nancy area under the British G.O.C., so that they might work with British bombing squadrons.

At present they are not inclined to put their squadrons with our wings but only to work with them in co-operation and not under them, but they are willing to use and work under us a common depot for repair work, &c. I think we should probably help them more by having them as an integral part of the British Forces; at the same time I hope to be able, with their co-operation, to get much more bombing carried out than our small programme will allow to begin with.

*5. Bombing policy from Nancy.*

It is intended to attack, with as large a force as is available, the big industrial centres on the Rhine and in its vicinity, in accordance with an organized plan. If the French do any long-distance bombing I am hopeful they will attack the targets we suggest.

It is also intended, when the weather is unfit for long-distance bombing, to attack nearer targets such as the big steel works near Briey, Saarbrucken, &c. This will be a development, on a larger scale, of work already successfully undertaken. It will be carried out under the orders of the French who will select, from time to time, the particular objective.

6. *Bombing from Dunkerque.*

It is also intended to bomb the submarine docks, &c., at Bruges and Zeebrugge. For this purpose there are already some Handley Page machines at Dunkerque which must be made up to two squadrons of full strength. These, with one D.H.9 (or 4) squadron will make a nucleus which by July should include the two H.P. squadrons and two D.H.9 (or 4) squadrons.

7. *Bombing enemy aerodromes used for housing machines which attack London.*

These aerodromes are in the Ghent neighbourhood or elsewhere within reach of Belgium and two night-bombing squadrons with short-range machines (F.E.2b's) will carry on this work.

8. *Short-range bombing by Army machines.*

Finally, the short-range bombing squadrons, which work in immediate connexion with the Army, will continue to bomb lines of communication, head-quarters, ammunition dumps, railway stations, aerodromes, &c., as at present.

13th January 1918.

## APPENDIX VI

### MEMORANDUM ON BOMBING OPERATIONS

(*Forwarded by C.I.G.S., War Office to General Sir H. Wilson, British Military Representative, Supreme War Council, on 17th January, 1918: Based on the contents of Appendix V*)

1. The bombing of long-distance targets in Germany (as distinguished from short-distance bombing in the manufacturing districts of Lorraine) was undertaken by the British Air Force in accordance with the orders of the British Cabinet given in October last.

The French Authorities warned the British that the organization and preparation of new aviation centres in the Nancy neighbourhood (from which long-distance bombing must, in view of the distances to be traversed, and the range of our existing machines, proceed) would be a very serious undertaking, and this has proved to be the case.

Nevertheless, five new aerodromes are under construction by the British in the Nancy neighbourhood, and these are expected to be sufficient to

## THE BOMBING OF GERMANY

accommodate twenty-five squadrons by the beginning of May. In the meantime, long-distance bombing is being carried on from Ochey, near Toul, where ground has been temporarily lent to us by the French.

It is hoped to have accommodation for fifteen more squadrons ready by the beginning of July, and still further accommodation in the same area soon afterwards.

Aircraft Parks, a Repair Depot, and a Stores Depot for this southern area are also being hurried forward.

2. The British are at present carrying out bombing, whenever weather conditions permit, from the Ochey area by *one* Handley Page squadron consisting of seven machines, *one* F.E.2b squadron consisting of eighteen machines and *one* D.H.4 (Rolls-Royce) squadron also consisting of eighteen machines.

The Handley Page machines are night bombers; the F.E.2b machines are only appropriate for short-distance night flying; the D.H.4 machines are used in the daylight. Later the day bombing machines will be the D.H.9.

3. As the main object for which British bombing squadrons were ordered to this area by the War Cabinet was the long-distance bombing of Germany, these squadrons are used as far as possible for raids upon the industrial towns on and near the Rhine, such as Mannheim and Karlsruhe. But these long distances are only possible when weather conditions are exceptionally favourable, and these machines cannot be expected, even under very suitable conditions, to go farther than Mannheim.

When long-distance bombing is not possible, short-distance targets are attacked, such as Briey and Saarebruck, with a view to knocking out the enemy's steel manufactures.

The selection of these short-distance targets is left to the French so that the closest co-operation is secured between the Allies in this work.

The French do not undertake systematic bombing of the Rhine towns, though occasionally French machines go a long distance as a specific act of reprisal. If long-distance bombing is regularly undertaken by the French it is hoped that their selection of targets may be made in co-operation with the British so as to secure the most effective results in this case also.

4. Every effort is being made by the British to send more bombing squadrons to this neighbourhood as rapidly as possible for the bombing of the Rhine towns. Careful estimates have been made and additional squadrons should be actually at work as follows:

|  | *Night bombing* | *Day bombing* | | *Total* |
|---|---|---|---|---|
| By 30th April | 1 (H.P.) | 2 (1 D.H.4 and | 1 D.H.9) | 3 |
| ,, 31st May | 1 ( ,, ) | 4 (1 ,, | 3 ,, ) | 5 |
| ,, 30th June | 1 ( ,, ) | 9 (1 ,, | 8 ,, ) | 10 |
| ,, 31st July | 2 ( ,, ) | 13 (1 ,, | 12 ,, ) | 15 |
| ,, 31st Aug. | 5 ( ,, ) | 18 (1 ,, | 17 ,, ) | 23 |
| ,, 30th Sept. | 8 ( ,, ) | 20 (1 ,, | 19 ,, ) | 28 |
| ,, 31st Oct. | 11 ( ,, ) | 20 (1 ,, | 19 ,, ) | 31 |

5. The policy intended to be followed is to attack the important German towns systematically, having regard to weather conditions and the defensive arrangements of the enemy. It is intended to concentrate on one town for successive days and then to pass to several other towns, returning to the first town until the target is thoroughly destroyed, or at any rate until the morale of workmen is so shaken that output is seriously interfered with.

6. Before leaving France, General Trenchard interviewed General Foulois with a view to seeing whether the American Air Forces, when they arrive, would co-operate in long-distance bombing. His suggestions have also been communicated in writing to General Pershing, in order to ascertain whether the Americans would be disposed to put their earliest bombing squadrons into a joint organization, or whether there is any other way in which the Allies already engaged in bombing Germany can assist the American Air Forces.

7. British experience goes to show that a great deal of time may inevitably be occupied in creating a large striking force, but that once a nucleus is created with its necessary aerodromes, depots, stores, &c., the work can be carried on and the force regularly increased. In order that the bombing of Germany may have its maximum effect it is of the first importance to carry through systematically, in spite of losses and other adverse conditions, the plan which has been made.

Long-distance bombing will produce its maximum moral effect only if the visits are constantly repeated at short intervals, so as to produce in each area bombed a sustained anxiety. It is this recurrent bombing, as opposed to isolated and spasmodic attacks, which interrupts industrial production and undermines public confidence.

On the other hand, if the enemy were to succeed in interrupting the continuity of the British bombing operations, their achievements (as the Allies success against Zeppelins show) would be an immense encouragement to them which would operate like a military victory.

The Allies must therefore adopt a programme of bombing operations which, whenever the weather permits, must be constantly kept up and under which it can be assured that the heavy losses, which are bound to occur, can be instantly made good.

## APPENDIX VII

MEMORANDUM BY SIR WILLIAM WEIR, SECRETARY OF STATE FOR THE ROYAL AIR FORCE, ON THE RESPONSIBILITY AND CONDUCT OF THE AIR MINISTRY, MAY 1918

1. The Air Force duly came into existance on the 1st April 1918. On my taking office on the 1st May, no definition of general policy regulating its conduct had as yet been laid down, although certain principles had been agreed, defining the responsibility for the operations of Air Force Units allocated to the Navy and Army.

2. I desire now to submit for the approval of the War Cabinet the following observations setting forth what I conceive to be some of the main responsibilities and duties attaching to the Air Ministry at this stage of the War.

3. In the first place, I assume that the decision of the War Cabinet to constitute the Air Ministry together with the Air General Staff was based on the conviction that the possibilities of aerial warfare and their influence on the war might more efficiently be realized by the establishment of a single authority, than by the maintenance of independent forces attached to the Navy and Army, and in particular, that a rapid development of aerial forces, devoted to the interruption of German industrial effort and kindred objects, might be achieved to such an extent as substantially to contribute towards bringing about a definite demand for peace. To attain success in this policy, the Air Ministry must therefore be recognized as the authority on general air policy, except as regards operations forming part of naval or military operations. It must act with the advice of and in the closest liaison with the representatives of the Navy and Army as regards the needs of these two Forces, and accordingly the Air Staff must command the fullest knowledge of all methods of the utilization of aircraft and their effectiveness in practice.

4. Hitherto, practically all aerial operations and the utilization of all aerial resources have been carried out under the orders of naval or military commanders, and to an overwhelming extent such operations have been in fields of aerial activity directly associated with naval and military efforts. Artillery co-operation, ground target work, tactical bombing, photographic work, anti-aircraft machines carried by warships, seaplane carrier units, escort and anti-submarine patrols, naval reconnaissances by use of aeroplanes and seaplanes and the use of torpedo-carrying aeroplane work, represent definite functions of aerial work associated at present with Naval and Military operations.

5. On the other hand, long- and extreme-range bombing machines for operation by day and night, utilized against targets outside the range of machines designed for the above functions, involve for their efficient utilization operational considerations of a purely aerial character, and require for their conception and execution a large measure of freedom and independence from other military schemes.

6. One factor which continuously tends to complicate the true war policy arises from the ever-growing series of functions and classes of activity to which aircraft might be applied with success. This results in a series of continuously growing demands on the necessarily limited aerial resources of the country. Whether these resources are assessed in aeronautical material or in personnel, their rate of development is continuous, but as shown by experience, such growth will not equal the rate of demand, although fortunately, the development should take place at a greater rate than that of the enemy's aerial resources.

7. In these circumstances, I submit the following proposition for approval:

(i) The Air Ministry is responsible to the War Cabinet for securing that utilization of the available aerial forces of the country which will prove most effective in its results against the enemy, with the limitation that the operational control of the forces allocated to the Navy or the Army rests with those Departments respectively.

(ii) This responsibility involves the power of allocating the resources as between the Navy, Army, and Air Force, and, in the event of a conflict of opinion as to forecast allocation arising owing to the forecast resources being below the demand of the three Authorities, the Secretary of State for Air, after consultation with the First Lord of the Admiralty and Secretary of State for War, must be the Judge, subject to appeal to the War Cabinet. Once an allocation is made in forecast, it can only be altered by consent subject to timely reference to the War Cabinet for a ruling should emergency arise.

(iii) A programme of the establishments of different units and of their provisional development will after consultation with Admiralty and War Office be regularly submitted to the War Cabinet based on the following considerations of war policy:

(a) The most effective strategical results will be obtained in the different spheres of activity by continuous and progressive measures rather than by isolated and non-continuous operations, except where such operations are of a justifiably opportunist character, or alternatively, hold substantial and definite potentialities.

(b) Large permanent establishments, unless continuously reviewed, tend to tie up resources, encourage hidden reserves, and prevent mobility and temporary concentration when such might be invaluable.

(c) The allocation to the present Western Front Army approximates to the maximum demanded on a well considered programme extending to June 1919.

(d) The necessity of quickly strengthening the anti-submarine aircraft patrols is clearly recognized.

(e) The continuous bombing of German industrial centres presents very important possibilities, and valuable results may be achieved by the use of even a small force commencing to operate now, and by its rapid progressive development.

(f) It is to be recognized that the Air Forces operated independently by the Air Ministry are to a certain extent mobile, and their availability is elastic, e.g. if the Army or Navy has some very important and temporary objective to attack by aircraft, and their own permanent establishment is unable to deal with it, they will be in a position to apply to the Air Ministry, who may detach temporarily sufficient resources to carry out the operation desired, but such Forces will be under the control of the Air Ministry for re-allocation. Such specially detached Forces will usually be operated under the Admiralty or War Office, but, in exceptional circumstances, they may by agreement with the Admiralty or War Office be operated by the Air Ministry.

(Sgd.) W. WEIR.

23rd May 1918.

# APPENDIX VIII

## MEMORANDUM OF MARSHAL FOCH

G.Q.G.A.
14th September 1918.

Supreme Command of the Allied Armies,
  1st Section,
      3rd Aviation Bureau.
No. 3919/A.

## MEMORANDUM ON THE SUBJECT OF AN INDEPENDENT AIR FORCE

The British point of view is as follows: 'The function of the Air Service 'must be regarded from three aspects.

'(1) The Air Service as an Auxiliary to the Navy, and as such subordinated to the Naval Authorities.

'(2) As an auxiliary to the Armies in the Field, and as such subordinated to the Military Command.

'(3) As undertaking long-distance bombing raids, and in this capacity independent of the Army and Navy.'

The logical development of this last point of view requires:

(1) The organization of a *very powerful* Air Force to bomb Germany.

(2) The Independence of this Force with respect to the command of the Allied Armies.

Quite apart from all technical questions relating to the utilization of the Air Service, the British point of view cannot be accepted in its entirety.

### I. *Organization*

Since the Military power of the enemy is represented by his Army, and since the sole means we have for destroying his Army is our own Army, it is on our Army that we must concentrate all our efforts in order to render it as strong and well-armed as possible. In consequence, any combination tending to diminish the Army or hinder its development on the plea of organizing a new force capable of reducing the enemy, must be rejected.

Moreover, the Army in order to beat the enemy seeks to engage him in combat, and in combat it seeks to achieve victory by means of the concentration and numerical superiority of the forces it brings into action. This means that any force which does not take part in combat is lost to the Army.

From this once more emerges the error of detailing men capable of fighting to a force which is not organized to take part in combat, as in the course of fighting to destroy the works and means necessary to the existence of the enemy.

The Bombing Air Service is not exempt from this rule and its use in war is measured by the effect which it can produce on the forces utilized by the enemy in battle. It can, however, in quiet periods, act on the morale of the enemy people or against enemy industrial establishments, which is its secondary function.

# MEMORANDUM OF MARSHAL FOCH

Yet again, no more than the Artillery, the Armoured cars, &c., can the Air Service by itself constitute an Army. If it is developed to an inordinate extent, this must, in view of the necessarily limited resources, inevitably be to the detriment of the other arms, and in particular of the infantry still of paramount importance, and so reduce the value of the whole Army.

## II. *Independence*

For the same reasons, and the British memorandum admits the point, it is inadmissible that the Air Force in question should not participate directly in the fighting which if it is to be brought to a victorious conclusion demands the concentration of all available resources.

The question as to the best use to be made of this force at each moment as circumstances require is therefore ever-present, and the decision as to whether it shall engage in the fighting or in long-distance raiding cannot, it seems, be given by any but the Commander-in-Chief.

Moreover, the Eastern Base of operations is the only base suited at the present time for bombing raids into Germany owing to its situation with respect to the targets chosen, and General Trenchard has provided and organized in this region the necessary establishment for the accommodation of the greater part, if not the whole of his Forces. In consequence, should important operations be undertaken East of Verdun, the presence in this district of numerous squadrons independent of the Chief Command, occupying a large part of the available ground, and overcrowding the lines of communication, will render the accommodation of the units detailed to take part in the fighting almost impossible, particularly in view of the limited amount of ground suitable for aviation in this region.

Moreover the operation of 'independent' squadrons must not be allowed to hinder the work of the fighting units which must inevitably occur if the work of the two Air Services is not regulated by a single command.

Finally, whether as regards organization or utilization, it is impossible to conceive of the development or even the existence of one of the combatant forces being exempted from the authority of the Chief Command, which is responsible for the united action of all these forces.

FOCH.

## APPENDIX IX

### JOINT NOTE NO. 35. BOMBING AIR FORCE. ADDRESSED TO THE SUPREME WAR COUNCIL BY THE MILITARY REPRESENTATIVES

Supreme War Council,
    Military Representatives.               Versailles,
                                                                     3rd August 1918

To: THE SUPREME WAR COUNCIL.

After a perusal of the Minutes of the 3rd Session of the Inter-Allied Aviation Committee, the Military Representatives consider that it is expedient, as soon as possible, to be in a position to carry out powerful and intensive bombardment, from the air, on enemy territory, both with

## BOMBING AIR FORCE

the object of destroying military objectives and, in case of need, to execute reprisals on German towns. They are therefore of opinion:

1. That, as soon as Allied resources in men and material permit, an Inter-Allied Bombing Air Force should be constituted in the Western theatre of operations, composed of weight-carrying machines with a wide radius of action, this force to be entirely at the disposal of the General Commanding-in-Chief the Allied Armies in France, the Commander of the Force to be nominated by him after consultation with the Commanders-in-Chief of the various Allied Armies under his orders.

2. That it is expedient, in anticipation of the constitution of this Force, and without waiting for the Governments to decide whether or not negotiations should be begun with the enemy, or whether he should be summoned to cease the bombardment of Allied towns under penalty of reprisals, to begin at once the elaboration of a methodical plan for the bombardment of towns and industrial centres belonging to the enemy.

3. That should an Inter-Allied Bombing Air Force be created in any other theatre of operations, this force should likewise be exclusively subject to the authority of the General Commanding-in-Chief the Armies operating in that theatre.

A plan for bombardment would then be established for that theatre in the same manner as shown above for the Western theatre.

| BELIN | G. SACKVILLE WEST | di ROBILANT | TASKER H. BLISS |
|---|---|---|---|
| | Major-General, | | |
| Military Representative, French Section, Supreme War Council. | Military Representative, British Section, Supreme War Council. | Military Representative, Italian Section, Supreme War Council. | Military Representative, American Section, Supreme War Council |

## APPENDIX X

### THE BOMBARDMENT OF THE INTERIOR OF GERMANY

*A memorandum[1] drawn up by Marshal Foch and sent to M. Clemenceau on the 13th of September 1918*

The MARSHAL COMMANDING-IN-CHIEF     Q.G. 13. 9. 1918.
the ALLIED ARMIES.
HEAD-QUARTERS.
*1st Section.*     MARSHAL FOCH.
*3rd* AVIATION BUREAU.     *Commander-in-Chief of the Allied Armies.*
*No. 3875.*     *To the* PRESIDENT OF THE WAR COUNCIL.

In your letter No. 11.330.D, of August 13th, 1918, you asked me to draw up a general programme for the bombardment of the interior of Germany and to examine the conditions under which it would be possible to carry it out.

[1] Based on Joint Note No. 35. See text p. 108.

The following memorandum deals with these two questions. The following conclusions are drawn:

I. *Constitution of an inter-allied bombing force.*

(*a*) To group together under the command of General Trenchard, the British squadrons, French and Italian Caproni squadrons at present assembled in the East and thus to form the nucleus of an inter-allied force destined to deal with the interior of Germany.

To this nucleus the squadrons of the Trenchard programme will later be added as they are formed as also any available squadrons of the other allied nations possessing heavy weight-carrying machines with a wide range of action.

(*b*) The forces thus grouped will be at my disposal for the execution, according to my instructions, of the bombing programme accepted by the Supreme War Council. I reserve the right during operations to use the whole or a part of these forces in battle, whether by dividing them among the armies or by fixing objectives for them in connexion with operations either projected or in course of execution.

II. *Scheme for the utilization of the inter-allied force.*

From the programme mentioned in the above memorandum the following must be excised:

(1) The Lorraine–Luxembourg iron basin reserved for aeroplanes of limited weight-carrying capacity and range action which could be put at the disposal of the General commanding the Eastern Army Group.

(2) Also for the moment, the metallurgic group of the Ruhr district and the Rhine towns below Coblenz, owing to their distance from our lines and the present weakness of our resources in regard to the importance and number of objectives. The objectives of this last category will be attacked as soon as the inter-allied force has a sufficient number of long-range aeroplanes available, to enable it to attack in an effective manner.

The targets to be bombed by the inter-allied force will therefore be in order of urgency:

1. *Distant objectives.*

    (*a*) The centres of chemical industry:
        Ludwigshafen–Oppau
        Hoechst–Leverkussen
    (*b*) The industrial and commercial centres of:
        Frankfurt
        Mannheim
        Mainz
        Stuttgart
        Aix-la-Chapelle
        Coblenz
        Friedrichshafen
    (*c*) The shunting station of Aix-la-Chapelle.

2. *Near objectives.*
   (a) The industrial and commercial centres of:
       Karlsruhe
       Freiburg
       Saarbrucken and the railway stations connected with them.
   (b) The shunting stations of:
       Ehrang
       Karthaus
       Offenburg
       Remelfingen (near Saargemund)
       Saarbrucken
       Hainsbergen (near Strassburg)

In addition the General commanding the inter-allied force will be given free scope to attack enemy aerodromes with a view to paralysing German Air Service units liable to hinder his work.

The above scheme can only be put into execution after approbation of the different Governments concerned.

It is in conformity with the wishes of the Military representatives of the Supreme War Council, contained in the Collective Note No. 35. It is therefore desirable that this Note be approved by the French Government and that you should use your high authority to hasten the consent of the other Allied Governments.

As soon as this consent is notified to me, I shall issue the necessary instructions.

(Sd.) Foch.

### I. Object

To carry the war into Germany by attacking her
industry (Munition work)
commerce (Economic crisis)
population (Demoralization)

These bombing raids on the German population do not properly speaking constitute reprisals—this like poison gas is a means of warfare which was first used by the enemy and which we are therefore forced to use in our turn.

### II. Scheme for Aerial Bombing

#### A. *Conditions which the Scheme must fulfil*

Air raids do not achieve good results unless they are on a *large scale* frequently *repeated* and form part of a methodical scheme carried out with tenacity.

Therefore, objectives to be bombed must fulfil the following conditions:
   (a) They must be within attacking distance of the front. They must be attacked—not intermittently but continuously—not by a few specially trained men but by whole bombing groups.
   (b) They must be of sufficient size to present a target fairly easy to find

and to attack, and such as to make defence difficult and necessitate considerable means of defence.

(c) They must be sufficiently stationary for any displacement of the targets out of bombing range to be impossible, the bombing of industrial establishments only achieving important destructive results after some time.

(d) *They must be as small in number as is necessary for effective action to be taken on each one.*

At the same time they must be well scattered over different districts and at sufficiently different points from the line to enable the Command to take weather conditions into consideration in the daily choice of objectives to be bombed and also to divide up the defences of the enemy.

### B. *Choice of Objectives to form part of the Bombing Scheme*

The examination of the economic situation of Germany reveals the following principal facts:

1. *The delicate situation as concerns her iron mines.*

At least 75 per cent. of iron-ore necessary to German consumption comes from the *Lorraine–Luxembourg* iron-ore area comprised by *Luxembourg–Longwy–Longuyon–Conflans–Metz–Thionville*, which is entirely within the range of our bombing service and even within that of our old-type machines (Voisin).

To paralyse the Lorraine–Luxembourg area is to cut off at *its source the raw material indispensable to the metal industry of Germany and thus to cause the closing down of the greater part of the works of the Ruhr basin (especially Essen) as well as those of the Saar.*

This result can be obtained by destroying the seven railway stations on the periphery of the area and especially the three principal stations at *Thionville, Metz,* and *Bettembourg,* completed by attacks on factories and mining centres; the distance of these objectives from the front varies between 75 and 30 km.

The stocks of iron in the works of the metal industry being exceedingly small a considerable dislocation of traffic at the stations which serve the iron-ore area will have an immediate effect on the output of these works.

It is therefore possible to obtain important results on the iron-ore area with limited means.

2. The great concentration of the large groups of works of the iron industry which are necessarily erected in a coal or iron-ore area.

The chief of these groups are:

(a) The Ruhr basin or Rhine–Westphalian basin (Essen district).

This is by far the most important (57 per cent. of the total steel production of Germany in 1916).

It provides excellent and numerous targets such as large factories and large shunting stations, but they are at a distance ranging between 260 and 320 km. from the front and therefore it is impossible to attack them effectively at any rate this year.

(b) The group—Saar basin, Lorraine basin, and Luxembourg basin (25 per cent. of the total production of Germany in 1916).[1]

It is between 40 and 100 km. from the front and can therefore be attacked in a consecutive manner but presents a great number of targets.

In order to achieve a considerable reduction in the production of this area, considerable means are required, and even were such a reduction obtained, it would not prove a decisive blow to the metal industry, as the chief centre, the Ruhr basin, would still remain untouched.

The best way of affecting the German metal industry *at the present date* appears therefore to be by attacking:

(a) not finished or manufactured products distributed among the numerous districts where they are turned out,

(b) nor steel works, the chief centre of which is at present out of our reach (Ruhr basin), but the very source of production, i.e. *localized supplies of iron ore*, the transport of which takes place via certain *obligatory routes*.

The Germans have themselves shown the capital importance with regard to the war of the possession and free exploitation of iron-ore; the confidential memorandum published by German metallurgists (December 1917) is particularly significant on the subject.

3. *Shortage and dilapidated condition of rolling-stock.*

This shortage, revealed alike in official reports and the press, is the only cause of the *coal crisis* and consequently of the decreased output of a great number of factories; this crisis is extraordinary in a country where coal is plentiful and where considerable quantities of coal are left lying at the mine pits.

This shortage is so great that it was often stated during 1917 and the beginning of 1918 in the most authoritative German military centres that:

'If Germany succumbs it will be due to the lack of means of transport.'

Material necessarily becomes accumulated at the larger shunting stations where trains are made up and which are relatively few and it is consequently here that its destruction should be attempted with a view to increasing the gravity of the present situation.

4. *Great size and vulnerability of large chemical centres.*

The preparation of the chief products used in the manufacture of explosives and poisonous powders and gases is concentrated in a few very large factories.

Thus the concentrated manufacturing centres of these *basic products* should be attacked and not the numberless gunpowder, explosive, or poison gas factories which are distributed over the whole of Germany and nearly always placed far from other habitations.

The chief chemical works are situated as follows:
    Ludwigshafen–Oppau (over 16,000 workmen)
    Hoechst

[1] Most of the works in French Lorraine have been dismantled.

Leverkussen, with works of secondary importance at Elberfeld Biebrich near Mainz. The Casells works at Frankfurt and Griesheil near Frankfurt

Ludwigshafen–Oppau is 160 km. from the lines
Hoechst          230 ,,      ,,    ,,
Leverkussen      250 ,,      ,,    ,,

Of the three chief centres, only Ludwigshafen–Oppau can be seriously attacked at the present date. The two remaining centres must be attacked as soon as means are available.

The destruction of these three factories would doubtless considerably influence the course of the War.

5. Importance of manufacturing and commercial towns in the valley of the Rhine. Particularly Mannheim, Cologne, Dusseldorf, to which should be added Frankfurt and Stuttgart and, in the second place, Karlsruhe, Mainz, Coblenz, Friedrichshafen, Freiburg, Aix-la-Chapelle.

Important objectives the *number of which is limited*.

6. There still remain a great number of objectives of interest within reach of the Allies' bombing service, for instance, the large gunpowder factory at Rottweil, the Mauser factory at Oberndorff, Kaiserslautern, &c. These will constitute secondary targets when atmospheric conditions are not favourable to raids on main objectives and may also be attacked with a view to keeping the enemy uncertain as to which are our true objectives.

### III. Execution of the Scheme. Constitution of the Inter-Allied Force

#### A. *Means of Action*

It is more important than ever that the *Entente* should seek a military decision.

The *only* means of obtaining this decision is to employ all the *aerial and land forces* on the great principle of *concentration*.

Consequently:

*a.* The main object of the aerial forces should be action in battle, their action against the interior of Germany, although of undoubted importance from the point of view of economics and morale, can only be secondary. It should therefore only be undertaken when fighting requirements have been satisfied or during lulls in the fighting.

*b.* The aerial forces should be placed under the control of the High Command, in the same way as the Army, which will decide, according to the situation, the objective, and the resultant general distribution of the whole aerial forces.

Two periods must therefore be distinguished in carrying out the scheme of raids into GERMANY, namely, periods of operations and periods of stabilization.

1. *During the periods of operations.*

The requirements of the battlefield must first be satisfied, and as completely as possible. The importance of these requirements necessitates

# INTERIOR OF GERMANY

that the majority of the bombing Air Service should take part in these operations. (The scheme of operations drawn up for Spring 1919 provides for the intervention of 2,200 bombing aeroplanes in the fighting out of a total of 3,300, which will be in existence at this time according to the programmes drawn up.)

The forces still available for attacking the interior of Germany must thus be very much reduced; they will not, however, be entirely suppressed; once raids have been started, they should be carried on uninterruptedly, even with reduced means, in order that the population may be prevented from recovering, and that the means of defence (aeroplanes and guns) accumulated by the enemy to defend his country may be maintained far from the battle-field.

Consequently a proportion must be established between the aerial bombing forces intended for operations and of those detailed to continue our attack on Industrial Germany. This proportion can only be fixed by the Commander-in-Chief of the Allied Armies.

2. *During the period of stabilization.*

The attack on the interior of Germany becomes the most important task of our bombing Air Service.

It is, however, necessary to maintain sufficient aerial forces at the disposal of the Armies.

*a.* To ensure the supervision of the front.

*b.* To carry on aerial warfare by bombing aerodromes, especially those acting as bases for raids on PARIS and LONDON by enemy aeroplanes.

*c.* To bomb objectives, an attack on which appears opportune to the High Command.

The proportion also remains to be fixed of the bombing air service for the Armies and that intended for attacks on Industrial Germany; this proportion must also be established by the Commander-in-Chief of the Allied Armies in France.

On this basis, the Inter-Allied Air Force detailed for bombing the interior of Germany can be constituted henceforth as follows:

        British Independent Force in the East.
        French Bomber Group 2.
        Italian Bomber Group 18.
        Subsequent American Flights.

The Commander-in-Chief of the Allied Armies will increase or reduce the forces composing this unit according to circumstances, and as he may judge necessary.

Fighting units may be attached under the same conditions to the Inter-Allied Air Force in order to facilitate if necessary day raids.

### B. *Command*

The Inter-Allied Force must have a single command. Unity of command is indispensable in order that the preparation and execution of the scheme of continuity be carried out with the order and method necessary for success; it can be accomplished as follows:

Major-General Trenchard, now in Command of the Independent Force, to command the Inter-Allied Air Force engaged in bombing the interior of Germany. He will be assisted by a Staff consisting of:
 His present Staff Major.
 A French Officer.
 An Italian Officer.
 An American Officer (when American Flights are attached to the Inter-Allied Force).

General Trenchard will be immediately under the orders of the Marshal Commander-in-Chief of the Allied Armies—as regards operations.

From the territorial point of view and for general discipline he will conform to the instructions of the General-in-Command of the Eastern Army Group.

### C. *Distribution of Objectives*

1. Raids on the iron districts of Lorraine–Luxembourg can be carried out by aeroplanes having an average load and inconsiderable range.

For the greater homogeneity of the Inter-Allied Force engaged in bombing Germany, it is advisable to remove aeroplanes of this type and to detail aeroplanes carrying a heavy load and with a large range.

Under these conditions raids on the iron district must be eliminated from the plan of the Inter-Allied Force. These raids will be carried out by French aeroplanes having an average load, in the East under the direct order of the General-in-Command of the Eastern Army Group.

2. The remaining objectives are still too numerous and some of them too distant to permit of *all* being bombed immediately.

They must only be attacked as the means at the disposal of the Commander of the Inter-Allied Force permit of heavy and repeated action in each case.

The Commander will designate according to circumstances those objectives which it is important should be bombed.

The Commander of the Inter-Allied Force will divide these objectives between different units, taking into consideration the characteristics of the different types of aeroplanes and the circumstances at the moment.

Consequently the objectives to be attacked are the following:
 I. The Lorraine–Luxembourg iron basin.
   (*a*) Mines and Works—industrial agglomerations.
   (*b*) Stations controlling the traffic in ore, chiefly METZ, THIONVILLE, BETTEMBOURG, and EHRANG and KARTHAUS near Treves.
 II. Centres for the manufacture of chemical products.
   LUDWIGSHAFEN–OPPAU.
   HOECHST.
   LEVERKUSSEN.
 III. Large industrial and commercial centres.
   (*a*) MANNHEIM, STUTTGART, FRANKFURT, COLOGNE, DUSSELDORF.
   (*b*) MAINZ, KARLSRUHE, FREIBURG, COBLENZ, SAARBRUCKEN, FRIEDRICHSHAFEN, AIX-LA-CHAPELLE.

# INTERIOR OF GERMANY

The railway stations supplying these towns would be attacked at the same time.

IV. Large sorting stations of Western Germany.
   HAUSBERGEN near Strassburg.
   KALK NORD near Cologne.
   DUSSELDORF–DERENDORF.
   SAARBRUCKEN.
   REMELFINGEN near Sarrequemines.
   EHRANG and KARTHAUS near Treves.
   AIX-LA-CHAPELLE.
   OFFENBURG near Strassburg.
V. The principal metallurgical works of the RHENISH-WESTPHALIAN basin (particularly ESSEN) and of the SARRE basin.

In addition, he can add to the objectives of the scheme a certain number of secondary objectives (Rottweil, Oberndorf, Kaiserslautern, &c.) which can be bombed when atmospheric conditions do not permit the principal objectives to be reached. These attacks will have the additional advantages of keeping the enemy uncertain as to real objectives, of maintaining the population in a state of alarm, and of compelling the means of defence to be scattered. Nevertheless, they must be reduced to a minimum or dispersion of effort will result.

3. The Commander of the Inter-Allied Force in order to maintain aerial supremacy must by stern fighting oppose any enemy air forces who will attempt to hinder the action of his units, especially his day bombers. To this end he should attack enemy aerodromes and destroy them—an essential condition to the success of his operations. This destruction of aerodromes—for the same reason—must be undertaken by the E.A.G., according to the means at its disposal. In order that these raids be conducted methodically, a common plan for the two Aviation Forces is necessary.

This plan will be drawn up by General Trenchard.

### D. *Establishment of the Inter-Allied Force*

The question has already been discussed by the Inter-Allied Aviation Commission which, at a meeting on 31st May 1918, adopted the following resolution:

Each one of the Allies will have a zone assigned to it and will take possession of the aerodromes through the intermediary of the local French authorities, according to the usual rules. The Commander-in-Chief of the Allied Armies has the right to take possession of other positions, suitable for aerodromes, and even to commandeer aerodromes already established.

Outside these zones no aerodromes can be commandeered without permission from General Duval.

It is necessary to sanction and to state precisely this decision, taking into consideration the following points:

   *a.* The Commander-in-Chief of the Armies of the North and the North-East cannot give up the control of the zones which are allotted to the Allied Services, since they will be situated on territory of which he is in

Command and from whence he must allot all the establishments of the rear. He has therefore the right to grant or refuse the sites asked for by the Allied Air Services in their zones.

The necessity for providing and obtaining aerodrome requirements for numerous eventualities will furthermore oblige him to reserve for himself a certain number of aerodromes in the zones allotted to the Allied Air Services.

*b.* It is necessary to provide for the return to the local command of all aerodromes placed at the disposal of the Allied Air Service should the situation render necessary the occupation of these aerodromes by the Air Service units required to take part in the fighting.

In consequence, the arrangements of the aerodromes belonging to the bombing Air Service for raids on Germany could be as follows:

1. A special zone of aerodromes is allotted—in principle—to each of the Allied Air Service which form part of a force intended to carry out raids on Germany.

The limits of these zones for the British and American Bombing Services are indicated on a map enclosed. The Italian Air Service will be placed in the French Zone.

2. The Allied Air Service will choose their aerodromes within the zones allotted to them; this choice will be submitted to the approval of the Commander-in-Chief of the French Armies in the North and North-East.

3. A certain number of aerodromes situated outside their zones could, as far as he considers possible, be placed by the Commander-in-Chief at the disposal of the Allied Air Services, but they must be returned to him upon request at 24 hours' notice. The Allied Air Services are not to establish on these aerodromes units for night flying.

4. Within the zones of the British and American aerodromes the Commander-in-Chief of the French armies of the North and North-East may lay out and reserve a certain number of aerodromes in the case of active operations on his front. When these aerodromes are not in use they will be placed at the disposal of the Allied Air Services.

5. The Commander-in-Chief of the Allied Forces reserves to himself the right, should the situation render it necessary, to order the evacuation of the whole or of a part of the aerodromes occupied by the allied bombing units, and to put these aerodromes at the disposal of the General Commanding operations in the eastern sector, in order to accommodate the Air Service units which are to take part in the battle.

6. The Commander-in-Chief of the French Armies of the North and North-East is responsible for the organization of the whole of the allied bombing units in the eastern sector, the aerial night routes and the establishment of signals.

No special arrangements of lights or of night signals are to be established without his authorization.

# APPENDIX XI

## HEADS OF AGREEMENT AS TO THE CONSTITUTION OF THE INTER-ALLIED INDEPENDENT AIR FORCE

(*An agreement reached between the British and French Governments and transmitted, through the Supreme War Council, to the American and Italian Governments for approval.*)

1. *The object of the force.*

To carry war into Germany by attacking her industry, commerce, and population.

2. *The plan of campaign.*

Air raids must be on a large scale and repeated, forming part of a methodical plan and carried on with tenacity.

3. *Execution of the plan.*

The complete realization of this scheme is not to be undertaken until the imperative requirements of the fighting have been satisfied or during the intervals of the fighting.

It will, therefore, be possible to carry out this plan in two ways according to circumstances.

(*a*) *During periods of active operations.* The requirements of battle will have to be met first, thus reducing in varying proportions the strength of forces available for raids on the interior of Germany. The bombing action being begun will, however, have to be pursued *even with a reduced strength*.

(*b*) *During steady or quiet periods.* Bombing raids on the interior of Germany become the chief work of our bombing squadrons. Having satisfied the Air Service requirements of the Army, all available long-range aeroplanes will be free to take part in the raids.

4. *The establishment of the Inter-Allied Independent Air Force.*

This establishment will include Allied Flights of heavy weight carrying aeroplanes with a wide radius of action, and will probably be reinforced later by further available Allied Flights of the same type.

The Force will be placed under the command of General Trenchard, as Commander-in-Chief, assisted by a staff including besides the present staff, an officer of each of the Nations represented in the Bombing Force.

General Trenchard will be under the Supreme Command of Marshal Foch for operations.

5. The name of the Force shall be the Inter-Allied Independent Air Force.

3rd October 1918.

# APPENDIX

## STATISTICS OF WORK OF SQUADRONS OF THE INDEP

| Date. | Hours flown. Day. | Hours flown. Night. | Weight of bombs dropped (nearest ton). | Photographic reconnaissances. No. of plates exposed. | Wastage in D.H.4. M. | Wastage in D.H.4. W. | D.H.9. M. | D.H.9. W. | D. M |
|---|---|---|---|---|---|---|---|---|---|
| 1918 June (6th–30th) | 1,514 | 399 | 57 | 748 | 3 | 7 | 6 | 13 | .. |
| July | 1,768 | 767 | 88 | 845 | 2 | 6 | 13 | 22 | .. |
| August | 2,019 | 846 | 101 | 704 | 7 | 16 | 14 | 22 | .. |
| September | 1,605 | 761 | 179 | 946 | 4 | 14 | 16 | 20 | 6 |
| October | 1,828 | 629 | 98 | 584 | .. | 7 | 2 | 13 | 11 |
| November (1st–10th) | 661 | 109 | 20 | 87 | 2 | 1 | 3 | 4 | .. |
| TOTALS | 9,395 | 3,511 | 543 | 3,914 | 18 | 51 | 54 | 94 | 17 |
| | | | | | 69 | | 148 | | |

Note. (1) The figures given above show a wastag
(2) The total personnel battle casualties fo
Killed .
Wounded .
Missing .
The detail of personnel casualties and w

| Sqdn. | Personnel. K. | Personnel. W. | Personnel. M. | D.H |
|---|---|---|---|---|
| 45 | .. | .. | 1 | . |
| 55 | 13 | 11 | 36 | 1 |
| 97 | 3 | 1 | 9 | . |
| 99 | 6 | 16 | 42 | . |
| 100 | 1 | 3 | 5 | . |
| 104 | 5 | 24 | 66 | . |
| 110 | 1 | 4 | 34 | . |
| 115 | .. | 1 | 3 | . |
| 215 | .. | 2 | 30 | . |
| 216 | .. | 2 | 9 | . |
| TOTALS | 29 | 64 | 235 | 1 |

(3) Of the total weight of bombs dropped some 390 tons were
(4) The weight of bombs dropped on enemy aerodromes was sor
Railways, &c.).

# XII

...DENT FORCE, INCLUDING WASTAGE. JUNE–NOVEMBER 1918

...lanes (Missing and Wrecked).

| F.E.2b. | | H.P. | | Sopwith 'Camel'. | | Total. | | Remarks. |
| --- | --- | --- | --- | --- | --- | --- | --- | --- |
| M. | W. | M. | W. | M. | W. | M. | W. | |
| .. | 4 | .. | .. | .. | .. | 9 | 24 | Maj. Gen. H. M. Trenchard took over tactical command of the Independent Force R.A.F. on 6th June 1918. The strength at this date in squadrons was one D.H.4 (No. 55), two D.H.9 (Nos. 99 and 104), for day bombing, and one H.P. (Naval 'A' later No. 216) and one F.E.2b (No. 100) for night bombing. Total 5. |
| | | | | | | 33 | | |
| .. | 10 | 1 | 3 | .. | .. | 16 | 41 | |
| | | | | | | 57 | | |
| 1 | 5 | 5 | 11 | .. | .. | 27 | 54 | No. 97 (H.P.) Squadron arrived 9th August 1918. First raid 19th/20th August 1918. No. 215 (H.P.) Squadron arrived 19th August 1918. First raid 22nd/23rd August 1918. |
| | | | | | | 81 | | |
| .. | .. | 11 | 13 | .. | .. | 37 | 54 | No. 100 (F.E.2b) Squadron commenced to equip with H.P. aeroplanes on 13th August 1918. No. 115 (H.P.) Squadron arrived 31st August 1918. First raid 16th/17th September 1918. |
| | | | | | | 91 | | |
| .. | .. | 1 | 18 | .. | .. | 14 | 59 | No. 110 (D.H.9a) Squadron arrived 31st August 1918. First raid 14th September 1918. No. 45 (Sop. 'Camel') Squadron arrived 22nd September 1918. |
| | | | | | | 73 | | |
| .. | .. | .. | 6 | 1 | .. | 6 | 11 | |
| | | | | | | 17 | | |
| 1 | 19 | 18 | 51 | 1 | .. | 109 | 243 | |
| | 20 | | 69 | | 1 | | 352 | |

one aeroplane for every 1·54 tons of bombs dropped.
...s period were as follows:
    . . . 29
    . . . 64
    . . . 235
...e in missing aeroplanes, by Squadrons, is as follows:

| Wastage in missing aeroplanes. | | | | |
| --- | --- | --- | --- | --- |
| D.H.9. | D.H.9a. | F.E.2b. | H.P. | Sop. 'Camel'. |
| .. | .. | .. | .. | 1 |
| .. | .. | .. | .. | .. |
| .. | .. | .. | 3 | .. |
| 21 | .. | .. | .. | .. |
| .. | .. | 1 | 1 | .. |
| 33 | .. | .. | .. | .. |
| .. | 17 | .. | .. | .. |
| .. | .. | .. | 1 | .. |
| .. | .. | .. | 10 | .. |
| .. | .. | .. | 3 | .. |
| 54 | 17 | 1 | 18 | 1 |

...ped by night.
...o tons, leaving 323 tons for industrial targets (Factories, Blast furnaces,

## APPENDIX XIII

### INDUSTRIAL TARGETS BOMBED BY SQUADRONS OF THE 41ST WING, AND THE INDEPENDENT FORCE, ROYAL AIR FORCE. OCTOBER 1917–NOVEMBER 1918

(*Note.* Attacks on enemy aerodromes not included.)

| Date. | Objectives: (a) allotted, (b) actually bombed. | Sqdns. | No. of aeroplanes which bombed. (No. which set out in brackets.) | Remarks. |
|---|---|---|---|---|
| **1917** 17th Oct. | (a) and (b) Saarbrücken (Burbach works). | 55 | 8 (11) | Three D.H.4's returned with engine trouble. Works and houses damaged. Monetary damage, 17,000 marks. Casualties reported, 5 killed, 9 injured. |
| 21st Oct. | (a) and (b) Bous (factory W. of and railways). | 55 | 11 (12) | One D.H.4 returned with engine trouble. Formation attacked by E.A., of which four were driven down out of control. One of the D.H.4's forced-landed in enemy territory and the occupants were made prisoners. |
| 24/25th Oct. | (a) and (b) Saarbrücken (Burbach works). | Naval 'A' | 9 (9) | Two H.P.'s missing } Casualties reported at Saarbrücken, 6 injured. |
| 24/25th Oct. | (a) and (b) Railways between Falkenburg and Saarbrücken. | 100 | 14 (16) | Two F.E.2b's missing |
| 29/30th Oct. | (a) Volklingen (steel works and railways). (b) Saarbrücken (railways). | 100 | 3 (9) | Weather unfavourable, 6 pilots lost direction and returned with bombs. |
| 30th Oct. | (a) and (b) Pirmasens (tanning works). | 55 | 12 (12) | Houses damaged. Casualties reported, 1 killed, 4 injured. |
| 30/31st Oct. | (a) and (b) Volklingen (steel works and railways). | 100 | 12 (12) | Weather unfavourable, but objective reached and direct hits obtained on workshops. Monetary damage, 47,646 marks. |

| Date | Target | Force | Aircraft | Remarks |
|---|---|---|---|---|
| 30/31st Oct. | (a) Mannheim (factories). Alternative Zweibrücken (railways). (b) Raid abandoned. | Naval 'A' | .. (4) | Four H.P.'s attempted this raid but turned back owing to unfavourable weather conditions. One forced-landed in enemy territory and the occupants were made prisoners. |
| 1st Nov. | (a) and (b) Kaiserslautern (munition works). | 55 | 12 (12) | One formation of six D.H.4's attacked by E.A. and the D.H.4's were compelled to drop their bombs in order to manœuvre. One E.A. crashed. |
| 5th Dec. | (a) Mannheim (chemical works). Alternative, Dillingen (Dillinger works). | | | Weather unfavourable. |
| | (b) Zweibrücken (railways). | 55 | 6 (6) | Casualties reported at Zweibrücken, 4 injured. |
| | (b) Saarbrücken (factories). | 55 | 5 (6) | One D.H.4 returned with engine trouble. Direct hits obtained on stores and repair shops and many houses were damaged. |
| 6th Dec. | (a) and (b) Saarbrücken (Burbach works). | 55 | 11 (12) | Casualties reported at Saarbrücken, 5 killed, 7 injured. One D.H.4 returned with engine trouble. Houses damaged. |
| 11th Dec. | (a) and (b) Pirmasens (railways). | 55 | 7 (7) | Casualties reported, 1 killed, 3 injured. |
| 24th Dec. | (a) and (b) Mannheim (factories and railways). | 55 | 10 (12) | Slight military damage reported. Casualties, 1 killed. Two D.H.4's returned with engine trouble. One forced-landed in enemy territory and the occupants were made prisoners. Workshops damaged. Monetary damage, 34,000 marks. |
| *1918* 3/4th Jan. | (a) and (b) Maizières (factories and railways). | 100 | 2 (10) | Casualties reported, 2 killed, 12 injured. Owing to unfavourable weather conditions only 2 F.E.2b's reached and bombed the target. |
| 4/5th Jan. | Ditto | 100 | 8 (9) | One F.E.2b returned with engine trouble. |
| 5/6th Jan. | (a) and (b) Conflans (railways). | 100 | 6 (9) | Two F.E.2b's returned with engine trouble, one lost direction and returned with bombs. |
| 5/6th Jan. | (a) Thionville (railways). Alternative, Maizières (factories and railways). (b) Courcelles (railway junction). | Naval 'A' | 1 (2) | One H.P. returned with engine trouble. Thick ground mist obscured target. |
| 14th Jan. | (a) and (b) Karlsruhe (factories and railways). | 55 | 12 (12) | Considerable damage caused to workshops. Monetary damage, 100,000 marks. |

| Date. | Objectives: (a) allotted, (b) actually bombed. | Sqdns. | No. of aeroplanes which bombed. (No. which set out in brackets.) | Remarks. |
|---|---|---|---|---|
| 1918 | | | | |
| 14/15th Jan. | (a) and (b) Diedenhofen (Thionville) (steel works and railways). | 100 | 9 (11) | Two F.E.2b's returned with engine trouble. |
| 16/17th Jan. | (a) Thionville (railways). (b) Bensdorf (railways). | 100 | 2 (6) | Four F.E.2b's returned owing to unfavourable weather conditions. |
| 21/22nd Jan. | (a) and (b) Diedenhofen (Thionville) (steel works) and Bensdorf (railways). | 100 | 12 (17) | Five F.E.2b's returned with engine trouble. One F.E.2b missing. |
| 21/22nd Jan. | (a) Treves (barracks and railways). Alternative, Diedenhofen (Thionville). (b) Arnaville (railway bridge near). | 16 (Naval) | 1 (2) | One H.P. returned owing to unfavourable weather conditions. The other H.P. encountered very heavy A.A. fire. |
| 24/25th Jan. | (a) and (b) Treves (barracks and railways) and Diedenhofen (Thionville) (steel works). | 100 | 11 (12) | One F.E.2b returned with engine trouble. One F.E.2b missing. |
| 24/25th Jan. | (a) and (b) Mannheim (chemical works) and Diedenhofen (Thionville) (factories and railways). | 16 (Naval) | 2 (3) | One H.P. returned with engine trouble. |
| 27th Jan. | (a) and (b) Treves (barracks and railways). | 55 | 6 (12) | Six D.H.4's returned owing to unfavourable weather conditions. |
| 9/10th Feb. | (a) and (b) Courcelles (railways). | 100 | 6 (6) | One F.E.2b missing. |

| Date | Target | Squadron | | Remarks |
|---|---|---|---|---|
| 12th Feb. | (a) and (b) Offenburg (railways and barracks). | 55 | 12 (12) | One F.E.2b and one H.P. returned with engine trouble. |
| 16/17th Feb. | (a) and (b) Conflans (railways). | 100 and 16 (Naval) | 5 (7) | |
| 17th Feb. | (a) Mannheim (chemical works). (b) Raid abandoned. | 55 | .. (12) | Rain and clouds prevented formation from seeing the ground, so returned with bombs. |
| 17/18th Feb. | (a) and (b) Conflans (railways). | 100 | 6 (6) | |
| 18th Feb. | (a) and (b) Treves (factories and railways). | 55 | 5 (5) | |
| 18th Feb. | (a) and (b) Thionville (factories and railways). | 55 | 4 (6) | Two D.H.4's returned with engine trouble. |
| 18/19th Feb. | (a) and (b) Treves (barracks and railways). | 100 | 8 (8) | One F.E.2b missing. Direct hits obtained on central station which caused a large fire. |
| 18/19th Feb. | (a) and (b) Thionville (factories and railways). | 100 | 2 (2) | Direct hits obtained on blast furnaces and gas works. |
| 19th Feb. | (a) Mannheim (chemical works). Alternative, Landau (barracks and railways). (b) Treves (workshops and railways). | 55 | 11 (12) | One D.H.4 returned with engine trouble. As the Valley of the Rhine was covered in thick mist, leader decided to bomb Treves. Formation attacked by E.A. and one D.H.4 failed to return. Barracks, coal depots, workshops, and houses damaged. Casualties reported, 2 killed (1 soldier). |
| 19/20th Feb. | (a) and (b) Thionville (factories and railways). | 16 (Naval) | 2 (2) | |
| 20th Feb. | (a) Mannheim (chemical works). Alternative, Kaiserslautern (factories and railways). (b) Pirmasens (factories and railways). | 55 | 8 (10) | Two D.H.4's returned with engine trouble. Thick mist in Rhine Valley. Houses damaged and many civilians injured. |
| 26/27th Feb. | (a) and (b) Treves (barracks and railways). | 16 (Naval) | 1 (2) | One H.P. returned with engine trouble. |
| 9th March | (a) and (b) Mainz (factories, barracks and railways). | 55 | 10 (12) | Two D.H.4's returned with engine trouble. |

| Date. | Objectives:<br>(a) allotted,<br>(b) actually bombed. | Sqdns. | No. of aeroplanes which bombed. (No. which set out in brackets.) | Remarks. |
|---|---|---|---|---|
| 1918<br>10th March | (a) and (b) Stuttgart (Daimler works, factories, and railways). | 55 | 11 (12) | One D.H.4 returned with engine trouble. Formation attacked by E.A. after objective had been bombed and one E.A. driven down out of control. One D.H.4 forced-landed and the occupants were made prisoners. Casualties reported at Stuttgart and district, 5 injured. Monetary damage estimated at 80,000 marks. |
| 12th March | (a) and (b) Coblenz (factories, barracks, and railways). | 55 | 9 (12) | Three D.H.4's returned with engine trouble. Formation attacked by E.A. and one E.A. driven down out of control. Considerable damage to town, and a direct hit obtained on the barracks. Casualties reported, 9 killed (4 soldiers) 61 injured (12 soldiers). |
| 13th March | (a) and (b) Freiburg (factories, barracks, and railways). | 55 | 8 (9) | One D.H.4 returned with engine trouble. On leaving objective formation attacked by E.A. and three of the enemy aeroplanes were driven down out of control. Three D.H.4's failed to return. Considerable damage was caused to public buildings and houses. Casualties reported, 5 injured. |
| 16th March | (a) Mannheim (chemical works). Alternative, Kaiserslautern (factories and railways).<br>(b) Zweibrücken (barracks and railways). | 55 | 7 (10) | Two D.H.4's returned with engine trouble. Formation attacked by E.A. before reaching objective, and one D.H.4 returned with damaged petrol tank, remainder of formation then proceeded to Zweibrücken. |
| 17th March | (a) and (b) Kaiserslautern (factories and railways). | 55 | 9 (10) | One D.H.4 returned with engine trouble. Many houses and workshops damaged. Monetary damage estimated at 124,000 marks. Reported many people killed and injured. |

| | | | |
|---|---|---|---|
| 18th March | (a) and (b) Mannheim (chemical works). | 55 | 9 (10) | One D.H.4 returned with engine trouble. Formation was attacked over objective and two E.A. were driven down out of control. Direct hits obtained on the works and fires started. Casualties reported, 4 killed, 10 injured. One H.P. returned owing to thick mist. |
| 23/24th March | (a) and (b) Conz (railways and bridges). | 16 (Naval) | 1 (2) | |
| 24th March | (a) and (b) Mannheim (chemical works). | 55 | 12 (12) | Over the objective the D.H.4's were attacked by a large formation of E.A. Two of the enemy aeroplanes crashed and four were driven down out of control. Two D.H.4's missing. |
| 24/25th March | (a) and (b) Thionville (railways). | 100 | 5 (6) | One F.E.2b returned with engine trouble. |
| 24/25th March | (a) and (b) Metz–Sablon (railways). | 100 | 12 (14) | Two F.E.2b's returned with engine trouble. Direct hits obtained on munition goods train which exploded and was burnt out. Much damage was caused to tracks, buildings, and the gasometer. Casualties reported, 6 killed, 2 injured. |
| 24/25th March | (a) Cologne and Luxembourg (railways). (b) Cologne, Luxembourg, and Courcelles (railways). | 16 (Naval) | 3 (3) | Owing to engine trouble one of the H.P.'s dropped his bombs on Courcelles. |
| 27th March | (a) and (b) Metz–Sablon (railways). | 55 | 11 (12) | One D.H.4 returned with engine trouble. |
| 28th March | (a) and (b) Luxembourg (railways). | 55 | 12 (12) | |
| 5th April | (a) Spa, Chateaux near (German G.H.Q.). (b) Luxembourg (station and railways). | 55 | 12 (12) | Leader considered it impossible to reach Spa. Weather cloudy. Strong S.W. wind. One D.H.4 failed to return. |
| 11th April | (a) Spa, Chateaux near (German G.H.Q.). Alternative, Luxembourg, Thionville, or Metz (railways). (b) Luxembourg (railways). | 55 | 11 (12) | One D.H.4 returned with engine trouble. Owing to thick clouds and high wind Luxembourg was chosen as the objective. |

| Date. | Objectives: (a) allotted, (b) actually bombed. | Sqdns. | No. of aeroplanes which bombed. (No. which set out in brackets.) | Remarks. |
|---|---|---|---|---|
| *1918* 12th April | (a) Spa, Chateau near (German G.H.Q.). (Alternative, Thionville or Metz–Sablon (railways). (b) Metz–Sablon (railways). | 55 | 11 (12) | One D.H.4 returned with engine trouble. Spa could not be reached owing to unfavourable weather conditions. Thionville was covered by clouds, so formation turned back and bombed Metz-Sablon. |
| 12/13th April 12/13th April | (a) and (b) Juniville (railways). (a) and (b) Juniville and Amagne–Lucquy (railways). | 100 216 | 6 (8) 3 (3) | One F.E.2b returned owing to bad visibility; one forced-landed. |
| 19/20th April | (a) and (b) Juniville (railways). | 100 | 6 (10) | One F.E.2b returned with engine trouble, three returned owing to bad visibility. |
| 19/20th April | (a) and (b) Juniville and Bethenville (railways). | 216 | 3 (3) | |
| 20/21st April | (a) and (b) Chaulnes and Roye (railways). | 100 | 7 (18) | Eleven F.E.2b's returned owing to bad visibility. |
| 20/21st April 21/22nd April | (a) and (b) ditto. (a) and (b) Amagne–Lucquy and Juniville (railways). | 216 100 | 2 (3) 9 (9) | One H.P. returned owing to bad visibility. |
| 2nd May 2/3rd May | (a) and (b) Thionville (railways). (a) Mohon (railways). (b) Amagne–Lucquy, Juniville and Warneville (railways). | 55 100 | 11 (12) 9 (18) | One D.H.4 returned, lost formation. Nine F.E.2b's returned owing to bad visibility. |
| 3rd May | (a) and (b) Thionville (railways). | 55 | 12 (12) | Serious damage to goods station, tracks and rolling stock. Monetary damage 25,000 marks. |
| 3/4th May | (a) Mohon (railways). | 100 | 9 (18) | Nine F.E.2b's returned owing to bad visibility. |

| Date | Objective | | | Remarks |
|---|---|---|---|---|
| 4th May | (b) Juniville, Asfeld, and Amagne–Lucquy (railways). (a) Thionville (railways). (b) Raid abandoned. | 55 | .. (12) | All D.H.4's forced to return owing to thick clouds. |
| 15th May | (a) and (b) Thionville (railways). | 55 | 12 (12) | Railway works hit. Monetary damage, 19,000 marks. |
| 16th May | (a) and (b) Saarbrücken (factories and railways). | 55 | 12 (12) | Severe fighting took place over objective, three E.A. driven down out of control. One D.H.4 shot down in flames. Occupants killed. Direct hits obtained on workshops, sidings, trains, and signalling apparatus. Casualties reported, 12 soldiers killed, 49 injured. |
| 17th May | (a) and (b) Metz–Sablon (railways). | 55 | 12 (12) | Direct hits obtained on main station, express train, and goods sheds. Casualties reported, 11 officers killed, 46 other ranks wounded. In addition some 30 soldiers and civilians killed and injured. |
| 17/18th May | (a) and (b) Thionville (railways). | 216 | 1 (2) | One H.P. returned with engine trouble. |
| 17/18th May | (a) and (b) ditto and Metz–Sablon (railways). | 100 | 10 (13) | One F.E.2b returned with engine trouble, one forced-landed, one missing. Considerable damage caused to town of Thionville during these raids. Reported, 35 persons killed. |
| 18th May | (a) and (b) Cologne (railways, workshops). | 55 | 6 (6) | On leaving objective D.H.4's attacked by two formations of E.A. Two of the enemy aeroplanes were driven down out of control. Serious damage to buildings in Cologne. Estimated casualties, 40 killed, 100 injured. Monetary damage estimated at 340,000 marks. |
| 20th May | (a) and (b) Mannheim (chemical works) or Landau (barracks and railways). | 55 | 10 (12) | Two D.H.4's returned with engine trouble. As Mannheim was covered in clouds, leader turned back and bombed alternative target. Casualties reported, 3 killed. |
| 20/21st May | (a) and (b) Coblenz, Thionville, Metz–Sablon (barracks and railways). | 216 | 6 (6) | Military damage reported. |
| 20/21st May | (a) and (b) Coblenz (barracks and railways) or Thionville (railways). | 100 | 11 (14) | Three F.E.2b's returned with engine trouble. Owing to bad visibility alternative objective bombed. |

| Date. | Objectives: (a) allotted, (b) actually bombed. | Sqdns. | No. of aeroplanes which bombed. (No. which set out in brackets.) | Remarks. |
|---|---|---|---|---|
| 1918 21st May | (a) Liége (railway triangle S. of, at Kinkempois). (b) Charleroi and Namur (railways). | 55 | 11 (12) | One D.H.4 returned with engine trouble. One D.H.4 failed to return. Owing to thick banks of fog leader tried to make Liége by compass, but owing to a strong N.E. wind, however, the formations arrived over Charleroi and Namur and bombed the railways at these places. |
| 21st May | (a) and (b) Metz-Sablon railways. | 99 | 6 (6) | |
| 21/22nd May | (a) and (b) Mannheim (chemical works), Thionville and Karthaus (railways). | 216 | 7 (7) | One H.P. forced-landed in enemy territory due to enemy A.A. fire and occupants taken prisoners. Direct hit obtained on the Oppau Works, which destroyed a gas main and caused a stoppage of work for two days. Railway workshops at Karthaus hit and engines damaged. |
| 21/22nd May | (a) and (b) Saarbrücken and Thionville (railways and factories). | 100 | 13 (16) | Three F.E.2b's forced-landed in our lines with engine trouble. Direct hits obtained on railway tracks and sidings at Saarbrücken, which caused serious dislocation of traffic. |
| 22nd May | (a) and (b) Liége (railway, triangle S. of, at Kinkempois). | 55 | 11 (12) | One D.H.4 returned with engine trouble. |
| 22nd May | (a) and (b) Metz-Sablon (railways). | 99 | 6 (6) | Direct hits obtained on railway stores. |
| 22/23rd May | (a) and (b) Kreuzwald (electric power station). | 100 | 11 (14) | Three F.E.2b's returned with engine trouble. |
| 22/23rd May | (a) and (b) Kreuzwald (electric power station) and Mannheim (chemical works). | 216 | 5 (5) | |

| Date | Objective | Sqn | Tons dropped | Remarks |
|---|---|---|---|---|
| 23rd May | (a) and (b) Metz–Sablon (railways). | 99 | 11 (12) | One D.H.9 returned with engine trouble. Casualties reported in this area, 4 killed, 15 wounded. |
| 24th May | (a) and (b) Hagendingen (Thyssen blast furnaces and works). | 99 | 8 (12) | Three D.H.9's returned, lost formation; one returned with engine trouble. Formation attacked over the objective and one E.A. driven down out of control. |
| 27th May | (a) and (b) Bensdorf (railways). | 99 | 10 (12) | Two D.H.9's returned with engine trouble. Formation attacked by E.A. over objective. One of the D.H.9's failed to return. |
| 27/28th May | (a) and (b) Kreuswald (electric power station) and Metz–Sablon railways. | 100 | 12 (18) | Three F.E.2b's returned with engine trouble. Two returned owing to bad visibility. One F.E.2b forced-landed in enemy lines, pilot and observer escaped, however, and re-joined Squadron two days later. |
| 27/28th May | (a) and (b) Mannheim (chemical works) and Kreuswald (electric power station), Landau and Courcelles (railways). | 216 | 5 (8) | One H.P. returned with engine trouble, two owing to bad visibility. Two H.P.'s compelled to return with engine failure dropped bombs on the railways at Landau and Courcelles. |
| 28th May | (a) and (b) Bensdorf (railways). | 99 | 10 (12) | Two D.H.9's returned with engine trouble. |
| 28/29th May | (a) and (b) Metz–Sablon (railways). | 100 | 4 (5) | One F.E.2b returned with engine trouble. One F.E.2b failed to return. |
| 29th May | (a) and (b) Metz–Sablon (railways). | 99 | 6 (12) | Six D.H.9's returned with engine trouble. Formation attacked over the objective and one E.A. driven down out of control. |
| 29th May | (a) Mannheim (chemical works). Alternative, Landau (barracks and railways). (b) Thionville (railways). | 55 | 11 (12) | One D.H.4 crashed on taking-off. Owing to the high wind the leader of the formation decided to bomb the railways at Thionville. |
| 30th May | (a) and (b) Thionville (railways). | 55 | 12 (12) | Town of Thionville damaged. Direct hit obtained on hotel in which four officers were killed and six wounded. Damage also caused to station, tracks, and rolling-stock. |
| 30th May | (a) Thionville (railways). (b) Metz–Sablon (railways). | 99 | 4 (12) | Two D.H.9's returned with engine trouble, six returned owing to bad visibility. |

| Date. | Objectives: (a) allotted. (b) actually bombed. | Sqdns. | No. of aeroplanes which bombed. (No. which set out in brackets.) | Remarks. |
|---|---|---|---|---|
| *1918* 30th May | (a) Thionville (railways). (b) Raid abandoned. | 99 | .. (11) | Eleven D.H.9's attempted a second raid on Thionville, but all returned owing to weather conditions. |
| 30/31st May | (a) Thionville (railways). | 100 | 6 (7) | One F.E.2b returned with engine trouble. Visibility extremely bad. |
| 30/31st May | (a) and (b) Metz–Sablon (railways). (b) Thionville, Metz-Sablon, Courcelles, Esch, and Conz (railways). | 216 | 7 (7) | One H.P. lost its course owing to poor visibility and bombed the railways at Conz. |
| 31st May | (a) Mannheim (chemical works). Alternative, Neustadt (railways). (b) Karlsruhe (railways and workshops). | 55 | 10 (12) | One D.H.4 returned with engine trouble. Owing to thick clouds N. of Pirmasens the leader of the formation decided to bomb Karlsruhe. Over the objective the D.H.4's were attacked by an enemy formation, and one E.A. was driven down out of control. One of the D.H.4's was shot down in flames. Considerable damage caused to houses and workshops. Factory production stopped for an hour. Casualties reported, 4 killed, 74 injured. Monetary damage, 700,000 marks. |
| 31st May | (a) and (b) Metz–Sablon (railways). | 99 | 10 (11) | One D.H.9 returned with engine trouble. |
| 31st May/1st June | (a) and (b) Karthaus and Metz-Sablon (railways). | 216 | 4 (7) | Three H.P.'s returned owing to bad visibility. |
| 31st May/1st June | (a) and (b) Thionville (railways) and Metz (railways). | 100 | 11 (15) | Four F.E.2b's returned with engine trouble. Metz–Sablon was also bombed. |
| 1st June | (a) and (b) Karthaus (railways and workshops). | 55 | 10 (12) | One D.H.4 returned with engine trouble, one seen to fall to pieces near Metz. Casualties reported, 2 killed, 4 injured. |

| Date | Target | Load | Sorties (Bombs) | Remarks |
|---|---|---|---|---|
| 1st June | (a) and (b) Metz–Sablon (railways). | 99 | 6 (12) | Six D.H.9's returned with engine trouble. |
| 2nd June | (a) Thionville (railways).<br>(b) Metz–Sablon (barracks and railways). | 99 | 10 (12) | Two D.H. 9's returned with engine trouble. |
| 3rd June | (a) Düren (factories). Alternative, Liége (railways near).<br>(b) Luxembourg (railways). | 55 | 7 (12) | Five D.H.4's returned with engine trouble. Thick clouds made the original objective impossible; leader decided to bomb the railway triangle at Luxembourg, but observation considerably hampered by clouds. |
| 3/4th June | (a) and (b) Metz–Sablon (railways). | 100 | 2 (6) | Owing to unfavourable weather conditions the recall signal was given but two machines not having seen signal crossed lines and bombed objective. |
| 4th June | (a) Coblenz (factories, stations, and barracks). Alternative, Conz (stations and workshops).<br>(b) Treves and Conz (railways and workshops). | 55 | 12 (12) | Thick clouds prevented D.H.4's reaching original objective; alternative objective covered with clouds, but some of the machines managed to bomb through breaks in the clouds, others bombed Treves. |
| 4th June | (a) Thionville (railways). Alternative, Hagendingen (railways).<br>(b) Metz–Sablon (railways). | 99 | 10 (12) | Two D.H.9's returned with engine trouble. Weather conditions unfavourable; formation unable to reach objective. |
| 5/6th June | (a) and (b) Metz–Sablon (railways). | 100 | 7 (7) | |
| 5/6th June | (a) and (b) Metz–Sablon (railways) and Thionville (railways). | 216 | 7 (7) | |
| 6th June | (a) and (b) Coblenz (factories, station, and barracks). | 55 | 10 (12) | Two D.H.4's returned with engine trouble. |
| 6th June | (a) and (b) Thionville (railways). | 99 | 5 (11) | Six D.H.9's returned with engine trouble. Direct hits obtained on railway workshops. Casualties reported, 3 killed, 1 injured. Monetary damage, 20,000 marks. |

| Date. | Objectives: (a) allotted, (b) actually bombed. | Sqdns. | No. of aeroplanes which bombed. (No. which set out in brackets.) | Remarks. |
|---|---|---|---|---|
| *1918* 6/7th June | (a) and (b) Thionville and Metz (railways). | 216 | 6 (7) | One H.P. returned with engine trouble. |
| 6/7th June 7th June | (a) and (b) ditto. (a) and (b) Coblenz (factories, station, and barracks). (b) Conz (railway sidings). | 100 55 | 10 (10) 9 (12) | Three D.H.4's returned with engine trouble, the remainder, owing to clouds, were unable to reach objective. Reported considerable material damage caused. Casualties, 4 killed, 7 injured. |
| 7th June 8th June | (a) and (b) Thionville (railways). (a) and (b) Metz-Sablon (railways). | 99 104 | 4 (6) 10 (12) | Two D.H.9's returned with engine trouble. Two D.H.9's returned with engine trouble. Attacked by hostile formation over Metz and one E.A. driven down out of control. A.A. fire heavy and accurate over objective. This was the first raid carried out by this Squadron in this area. |
| 8th June | (a) Coblenz (factories, station, and barracks. Alternative, Karthaus (railways and workshops). (b) Thionville (station and sidings). | 55 | 11 (12) | One D.H.4 returned with engine trouble. Formation could not get farther than Thionville on account of clouds. |
| 8th June | (a) and (b) Hagendingen (factories and railways). | 99 | 9 (12) | Two D.H.9's returned with engine trouble, one returned with sick pilot. |
| 9th June | (a) and (b) Dillingen (factory and station). | 99 | 12 (12) | |
| 9th June | (a) and (b) Hagendingen (factories and railways). | 104 | 6 (12) | Five D.H.9's returned with engine trouble, one returned with sick observer. |

| Date | Target | Sqdn | Machines | Remarks |
|---|---|---|---|---|
| 12th June | (a) Dillingen (Dillinger Hutton works). | 99 | 12 (12) | Weather unfavourable, visibility very bad. |
| 12th June | (b) Metz–Sablon (railways). (a) Hagendingen (factories and railways). (b) Metz–Sablon (railways). | 104 | 11 (12) | One D.H.9 returned with engine trouble. Visibility very bad. |
| 13th June | (a) Coblenz (station and factories). (b) Treves (station and barracks). | 55 | 12 (12) | Thick clouds north of Treves prevented bombing of original objective, so leader decided to bomb Treves. Over Treves the D.H.4's were attacked by E.A. and in the heavy fighting which ensued two of the enemy aeroplanes were destroyed. One D.H.4 was shot down in flames. Considerable damage caused by direct hit on an iron foundry and other factories. Casualties reported, 4 killed, 8 injured. |
| 13th June | (a) and (b) Dillingen (Dillinger Hutton works). | 99 | 9 (12) | Three D.H.9's returned with engine trouble. |
| 13th June | (a) and (b) Hagendingen (factories and railways). | 104 | 8 (12) | Four D.H.9's returned with engine trouble. Formation attacked over objective and two E.A. driven down out of control. *Note.* During the raids on Hagendingen on the 8th, 9th, and 13th June, the factory was hit and the cement works badly damaged. |
| 23rd June | (a) and (b) Metz–Sablon (railways). | 55 | 11 (12) | One D.H.4 returned with engine trouble. |
| 23rd June | (a) and (b) ditto. | 99 | 12 (12) | |
| 23rd June | (a) and (b) ditto. | 104 | 10 (12) | One D.H.9 returned with engine trouble, one wrecked shortly after leaving aerodrome. |
| 23/24th June | (a) Buhl aerodrome. (b) Metz–Sablon (railways). | 100 | 12 (14) | One F.E.2b forced-landed in our lines, one returned with engine trouble. Weather unfavourable. |
| 24th June | (a) Mannheim (chemical works). Alternative, Kaiserslautern (factories and railways). | 55 | 10 (12) | Two D.H.4's returned with engine trouble. Thick clouds prevented formations reaching objectives. The two formations lost touch, one bombed Dillingen, the other Metz–Sablon. |

| Date. | Objectives: (a) allotted, (b) actually bombed. | Sqdns. | No. of aeroplanes which bombed. (No. which set out in brackets.) | Remarks. |
|---|---|---|---|---|
| 1918 24th June | (b) Dillingen (foundries and station) and Metz-Sablon (railways). | 99 | 10 (12) | Two D.H.9's returned with engine trouble. Owing to the clouds the formations were separated, one bombed the allotted objective, the other could not find the objective and bombed the factories and railways at Saarbrücken. |
| 24th June | (a) and (b) Dillingen (foundries and station) and Saarbrücken (railways). | 104 | 9 (12) | Three D.H.9's returned with engine trouble. Owing to thick clouds a compass course had to be steered and Saarbrücken was bombed. Over Saarbrücken the D.H.9's were attacked by hostile formation, and one E.A. was shot down in flames, another driven down out of control. Direct hits obtained on railways workshops and main station. Monetary damage 40,000 marks. |
| 25th June | (a) and (b) Saarbrücken (factories and railways). | 55 | 9 (12) | Two D.H.4's returned with engine trouble. One forced-landed in our lines. Formation attacked by E.A. over objective and a D.H.4 was shot down, the occupants being killed. |
| 25th June | (a) and (b) Offenburg (railways and barracks). | 99 | 11 (12) | One D.H.9 returned with sick pilot. Formation attacked over objective and two E.A. driven down out of control. One D.H.9 was forced to land in enemy territory. The objective was heavily bombed: slight casualties reported. Monetary damage, 30,000 marks. |
| 25th June | (a) and (b) Karlsruhe (munition factories). | 104 | 7 (12) | Four D.H.9's returned with engine trouble and one hit by A.A. fire was forced to return. The formation attacked by E.A. |

| Date | Target | Sqn | Sorties | Remarks |
|---|---|---|---|---|
| 25/26th June | (a) and (b) Metz–Sablon (railways), and Ars (railway station). | 216 | 5 (6) | and two of the enemy aeroplanes driven down out of control. One of the D.H.9's was forced to land in enemy territory. Many houses and workshops damaged. Casualties reported, one killed. Monetary damage, 250,000 marks. |
| 26th June | (a) and (b) Karlsruhe (factories and railways). | 55 | 11 (12) | One H.P. returned with engine trouble. One H.P. missed objective and bombed Ars station. Considerable damage caused to the Metz–Sablon tracks. |
| 26th June | (a) and (b) ditto. | 99 | 6 (12) | One D.H.4 returned with sick pilot. One D.H.4 forced-landed in enemy territory. |
| 26th June | (a) and (b) ditto. | 104 | 3 (12) | Five D.H.9's returned with engine trouble, one with sick observer. |
| 26/27th June | (a) and (b) Mannheim (chemical works) and Saarbrücken (factories and sidings). | 216 | 3 (4) | Three D.H.9's returned with engine trouble; four failed to rendezvous correctly and returned; one pilot lost his way, flew over Swiss territory, was fired on by Swiss A.A. defences and eventually crashed, pilot and observer unhurt; one other D.H.9 missing. |
| 27th June | (a) and (b) Thionville (station and workshops). | 55 | 11 (12) | One H.P. unable to reach the allotted objective owing to engine trouble, bombed Boulay aerodrome. Considerable material damage reported at Mannheim and Saarbrücken. Casualties reported, one killed. |
| 27th June | (a) and (b) ditto. | 99 | 11 (12) | One D.H.4 forced-landed in our lines with engine trouble. |
| 27th June | (a) and (b) ditto. | 104 | 5 (6) | One D.H.9 returned with engine trouble. |
| 27/28th June | (a) and (b) Metz–Sablon. | 100 | 5 (9) | One D.H.9 returned with engine trouble. *Note.* The formations of Nos. 55, 99 and 104 Sqdns. were attacked over the objective and three E.A. were destroyed and one driven down out of control. One D.H.9 (99 Squadron) failed to return. Four F.E.2b's lost direction owing to thick mist and returned. Boulay aerodrome also bombed. |
| 29th June | (a) and (b) Mannheim (chemical works). | 55 | 10 (11) | One D.H.4 returned with engine trouble. Over the objective D.H.4's attacked and two E.A. were driven down out of con- |

| Date. | Objectives: (a) allotted, (b) actually bombed. | Sqdns. | No. of aeroplanes which bombed. (No. which set out in brackets.) | Remarks. |
|---|---|---|---|---|
| 1918 29th June (cont.) | | | | trol. Damage caused to buildings. Casualties reported, 5 killed, 16 injured. Monetary damage, 151,000 marks. |
| 29/30th June | (a) and (b) Mannheim (chemical works), Thionville and Metz (railways). | 216 | 6 (7) | One H.P. returned with engine trouble. Only one H.P. reached and bombed the objective, the others attacked Thionville and Metz. |
| 30th June | (a) and (b) Landau (barracks and station). | 104 | 9 (11) | Two D.H.9's returned with engine trouble. Formation attacked on the way out and home, two E.A. were crashed and two driven down out of control. One D.H.9 failed to return. |
| 30th June/1st July | (a) and (b) Mannheim (chemical works), Thionville (railway workshops). | 216 | 7 (8) | One H.P. returned with engine trouble. Towns in the vicinity of these objectives were also bombed. |
| 1st July | (a) and (b) Karthaus and Treves (railways and workshops). | 55 | 11 (12) | One D.H.4 returned with engine trouble. |
| 1st July | (a) and (b) Karthaus (railways and workshops). | 99 | 6 (10) | One D.H.9 returned with engine trouble, two with sick pilots and observers, one missed formation. |
| 1st July | (a) Karthaus (railways and workshops). Alternative, Thionville (railways). (b) Metz-Sablon (railways). | 104 | 4 (10) | Two D.H.9's returned with engine trouble, one was wrecked on taking-off and one forced-landed in our lines. The formation had been attacked near Conflans and again heavily attacked over Metz. They destroyed one E.A. Two D.H.9's failed to return. The leader deemed it inadvisable to proceed to the allotted objective with only four D.H.9's. |
| 1/2nd July | (a) and (b) Mannheim (chemical works) and Thionville (railways and workshops). | 216 | 5 (6) | One H.P. returned with engine trouble. Houses damaged. Casualties reported, 1 killed, 3 injured. |

| | | | | |
|---|---|---|---|---|
| 2nd July | (a) and (b) Coblenz (railway sidings). | 55 | 9 (12) | Three D.H.4's returned with engine trouble. Some material damage was caused and two people were injured. |
| 2nd July | (a) Karthaus (station and bridges). (b) Treves. | 99 | 6 (9) | One D.H.9 returned with engine trouble, one with sick observer, one missed formation. Formation attacked over Treves, one E.A. crashed and one driven down out of control. |
| 5th July | (a) and (b) Coblenz (barracks and railways). | 55 | 11 (12) | One D.H.4 forced-landed in our lines. Considerable damage caused to buildings in the town. Casualties reported, 4 injured. |
| 5th July | (a) Kaiserslautern (factories and station). | 99 | 6 (6) | Formation attacked over objective and one E.A. destroyed. No bursts observed owing to the fighting, but some material damage in Saarbrücken was later reported. |
| 5th July | (a) Kaiserslautern (factories and station). (b) Barbas village S. of Blamont. | 104 | 4 (6) | Two D.H.9's returned with engine trouble. When over the lines a mechanical defect occurred to the leader's machine and he dropped his bombs on Barbas, the other pilots not realizing the situation also dropped their bombs. |
| 6th July | (a) Düren (explosive factory). (b) Metz–Sablon (railways). | 55 | 12 (12) | The first formation, owing to the leader's D.H.4 having had its radiator shot through, bombed Metz. The second formation, owing to darkness, also bombed Metz. |
| 6th July | (a) Düren (explosive factory). (b) Metz–Sablon (railways). | 99 | 6 (6) | Weather unfavourable. |
| 6th July | (a) Kaiserslautern (factories and station). (b) Metz–Sablon. | 104 | 6 (6) | Ditto. |
| 6/7th July | (a) and (b) Saarburg (railway station). | 100 | 7 (12) | Two F.E.2b's returned with engine trouble, three lost direction and returned. |
| 6/7th July | (a) Mannheim (chemical works). Alternative, Thionville (railway and workshops). (b) Metz (railways). | 216 | 3 (8) | Five H.P.'s returned owing to unfavourable weather conditions. Thick mist and heavy clouds made it impossible to see any objectives. Bombs dropped on approximate position of railway junction at Metz. Reported later that damage was caused to tracks and rolling-stock. |
| 7th July | (a) and (b) Kaiserslautern (railways and factories). | 99 | 6 (6) | |

| Date. | Objectives: (a) allotted, (b) actually bombed. | Sqdns. | No. of aeroplanes which bombed. (No. which set out in brackets.) | Remarks. |
|---|---|---|---|---|
| 1918 7th July | (a) and (b) Kaiserslautern (railways and factories). | 104 | 5 (6) | One D.H.9 crashed in forced landing in our lines. Formation attacked on outward and return journey and one E.A. driven down out of control. Two D.H.9's failed to return, one being shot down near Pirmasens. Some material damage was caused during these raids on Kaiserslautern. Casualties reported, one killed, one injured. |
| 8th July | (a) Duren (explosives factories). (b) Luxembourg (railways). | 55 | 12 (12) | Encountering clouds the leader turned and bombed Luxembourg. |
| 11th July | (a) and (b) Offenburg (railways). | 55 | 12 (12) | Two D.H.4's returned with engine trouble. Heavy clouds prevented the formation reaching objective allotted. |
| 12th July | (a) Offenburg (railways). (b) Saarburg (railways E. of) | 55 | 10 (12) | |
| 15th July | (a) and (b) Offenburg (railways). | 55 | 11 (12) | One D.H.4 returned with engine trouble. Material damage caused. Casualties reported, one killed, one injured. |
| 16th July | (a) Stuttgart (Bosche & Daimler works). Alternative, Oberndorf (munition works). (b) Thionville (station and railway workshop). | 99 | 12 (12) | Thunderstorm to the north compelled leader to turn and bomb Thionville. |
| 16th July | (a) (as for 99 Sqdn.). | 55 | 6 (12) | Four D.H.4's missed direction owing to thick clouds, one crashed in taking off and one returned with sick observer. During the raids on Thionville direct hits were obtained on a munition train (some 15 trucks exploded), horse transport train, goods station, sheds, tracks, and sidings. Serious fires caused and goods station badly damaged. Estimated casualties, |
| 16th July | (b) Thionville (station and railway workshops). | | | |

| Date | Target | | | Notes |
|---|---|---|---|---|
| 16/17th July | (a) and (b) Saarbrücken (Burbach works). | 216 | 3 (7) | 83 officers and other ranks killed and wounded. Ten civilians killed. Four H.P.'s returned with engine trouble. Five bombs fell in the centre of the town causing heavy material damage. |
| 16/17th July | (a) and (b) Hagendingen (blast furnaces) and Han (railway junction). | 100 | 10 (14) | Three F.E.2b's returned with engine trouble. One forced-landed in our lines. Han railway junction also bombed. At Hagendingen one bomb pierced the tunnel leading from the offices to the railway killing 9 people and seriously injuring 14. Supply store on Weiden Island set on fire. Several thousand tons of compressed straw and more than a hundred tons of coal destroyed. Office buildings destroyed. |
| 17th July | (a) Stuttgart (Bosche & Daimler works). Alternative, Oberndorf (munition works). (b) Thionville (station and workshops). | 99 | 6 (12) | Five D.H.9's returned with engine trouble, one lost formation and returned. |
| 17th July | (a) (as for 99 Sqdn.). (b) Thionville (station and workshops). | 55 | 11 (12) | One D.H.4 returned with engine trouble. |
| 18/19th July | (a) and (b) Mannheim (Benz or chemical works) and Saarbrücken (Burbach works). Heidelburg (railway station) and Wadgassen (blast furnaces). | 216 | 6 (8) | One H.P. forced-landed in our lines, one crashed just after taking-off. In addition to allotted objectives, Heidelburg and Wadgassen also bombed. |
| 19th July | (a) and (b) Oberndorf (munition works). | 55 | 8 (12) | Four D.H.4's returned with engine trouble. Slight material damage caused. |
| 19/20th July | (a) and (b) Mannheim (chemical works) and Saarbrücken (Burbach works). | 216 | 6 (6) | One H.P. failed to return. |
| 20th July | (a) and (b) Oberndorf (munition works). | 55 | 11 (12) | One D.H.4 returned with engine trouble. Formation attacked over objective, two E.A. were destroyed and one driven down |

| Date. | Objectives:<br>(a) allotted,<br>(b) actually bombed. | Sqdns. | No. of aeroplanes which bombed. (No. which set out in brackets.) | Remarks. |
|---|---|---|---|---|
| 1918<br>20th July (cont.) | (a) and (b) Offenburg (railways). | 99 | 10 (12) | out of control. Two D.H.4's were forced down in enemy territory. Many houses damaged in Oberndorf. Two D.H.9's returned with engine trouble. Formation attacked near objective and one E.A. driven down out of control. One D.H.9 was shot down in enemy territory. |
| 21/22nd July | (a) and (b) Mannheim (chemical works). Zweibrücken (factories). Lumes (railway sidings). | 216 | 4 (4) | |
| 22nd July | (a) Stuttgart (Bosche & Daimler works).<br>(b) Rottweil (munition factories). | 55 | 10 (12) | Two D.H.4's returned with engine trouble. Material damage caused. Strong winds prevented the D.H.4's from reaching Stuttgart. |
| 22nd July | (a) (as for 55 Sqdn.).<br>(b) Offenburg (railways). | 99 | 12 (12) | Owing to strong winds and mechanical trouble to leader's D.H.9, Offenburg was chosen as the objective. The formation was attacked by E.A., and one was destroyed and another driven down out of control. Many buildings damaged. Casualties reported, 1 killed, 2 injured. |
| 25/26th July | (a) Lumes (railways) and Stuttgart (Bosche & Daimler works). Alternative, Oberndorf (munition works).<br>(b) Pforzheim (factory and railways), Offenburg (railways). | 216 | 8 (8) | Unable to reach objectives allotted owing to heavy clouds in Rhine Valley. Many houses damaged in Offenburg neighbourhood. Casualties reported, 2 killed, 5 injured. |

| Date | Target | | | |
|---|---|---|---|---|
| 29/30th July | Balan Sedan (factory); also aerodromes at Boulay and Juvigny. | 216 | 7 (7) | The report states: 'Owing to the clouds machines were unable to reach Stuttgart and were forced to choose any objectives they were able to locate.' |
| 30th July | (a) Lumes (railways), Malmy aerodrome and Stuttgart (works). Alternative, Oberndorf (works).<br>(b) Offenburg, Rastatt, and Baden (railways), Sollingen wharves, and Malmy aerodrome. | | | |
| 30th July | (a) Stuttgart (Bosche & Daimler works).<br>(b) Offenburg (railways). | 55 | 10 (12) | One D.H.4 returned with engine trouble, one returned with sick observer. Owing to thick fog and low clouds the leader decided to make for Offenburg. |
| 30th July | (a) Stuttgart (Bosche & Daimler works).<br>(b) Lahr. | 99 | 8 (12) | Two D.H.9's returned with engine trouble, two lost formation and returned. Report states: 'Found low clouds over the country towards Stuttgart so bombed Lahr, there being no alternative target detailed.' Formation attacked by large formations of E.A. During the fighting three of the E.A. were destroyed and two driven down out of control. One D.H.9 was shot down over enemy territory. Considerable material damage to houses and machinery. Casualties reported, 3 injured. |
| 30/31st July | (a) and (b) Stuttgart (Bosche & Daimler works) and Hagenau (railway station). | 216 | 4 (7) | One H.P. returned with engine trouble, two owing to weather conditions. Material damage caused at Stuttgart. Hagenau also bombed. |
| 31st July | (a) and (b) Coblenz (factories, station, and barracks). | 55 | 10 (12) | Two D.H.4's returned with engine trouble. |
| 31st July | (a) Mainz (munition factories and depots). Alternative, Mannheim (Benz works).<br>(b) Saarbrücken. | 99 | 5 (12) | Three D.H.9's returned with engine trouble. On the way to the objective, formation attacked by E.A. over Saaralbe and one D.H.9 shot down. Shortly after the D.H.9's again attacked by large formations, estimated at 40, and three more D.H.9's were shot down. Leader then decided to bomb Saarbrücken |

| Date. | Objectives: (a) allotted, (b) actually bombed. | Sqdns. | No. of aeroplanes which bombed. (No. which set out in brackets.) | Remarks. |
|---|---|---|---|---|
| *1918* 31st July (*cont*). | | | | with the remaining five D.H.9's but was again attacked and a further three D.H.9's were shot down, making a total of seven D.H.9's lost over enemy territory. Two of the E.A. were claimed to have been driven down. The remaining two D.H.9's returned safely. Houses damaged in Saarbrücken. Casualties reported, 5 killed, 5 injured. |
| 31st July | (a) and (b) Saarbrücken (factories and sidings). | 104 | 12 (12) | |
| 1st Aug. | (a) Cologne (barracks and railways). Alternative, Coblenz (factories, station, and barracks). (b) Düren (factories and railways). | 55 | 11 (12) | One D.H.4 returned with engine trouble. Owing to clouds Cologne could not be seen so leader decided on Düren. Considerable damage caused to property. Casualties reported, 16 killed, 18 injured. Monetary damage estimated at 170,561 marks. |
| 1st Aug. | (a) Karthaus (railways and workshops). (b) Treves (workshops and sidings). | 104 | 10 (12) | Two D.H.9's returned with engine trouble. On reaching Karthaus the objective was covered in mist, so leader decided to bomb Treves. Over Treves the formation was attacked by E.A. and again near Boulay aerodrome. In the fighting, three of the enemy aeroplanes were destroyed and one driven down out of control. One D.H.9 failed to return. Considerable damage caused to houses and factories in Treves. Casualties reported, 1 killed, 4 injured. |
| 8th Aug. | (a) and (b) Wallingen (factory N.E. of Rombach). | 55 | 11 (12) | One D.H.4 returned with engine trouble. Direct hits obtained on factory. Rolling-mills, and blast furnaces damaged. Works |

| Date | Objective | | | Remarks |
|---|---|---|---|---|
| 11th Aug. | (a) Mannheim (Benz works). (b) Metz-Sablon (railway triangle). | 55 | 1 (10) | closed for 8 hours. Monetary damage estimated at 40,467 marks. Three D.H.4's returned with engine trouble, one returned with sick observer, five abandoned the raid. The remaining D.H.4 bombed Metz. |
| 11th Aug. | (a) Mannheim (Benz works). (b) Karlsruhe (railways). | 104 | 10 (12) | One D.H.4 returned having lost formation. On the way to objective formation continually attacked by E.A. Owing to thick clouds it was impossible to reach Mannheim and leader decided to bomb Karlsruhe. Over the objective formation again attacked and three E.A. were driven down out of control. One of the D.H.9's failed to return. Damage was caused to houses, workshops and an officers' prisoners of war camp. Monetary damage, 130,000 marks. |
| 12th Aug. | (a) and (b) Frankfurt (factories and railways). | 55 | 12 (12) | Formation attacked on leaving the objective and two E.A. were destroyed and two driven down out of control. Bombs fell in the centre of the town causing considerable damage to houses and factories. Casualties reported, 16 killed and 22 injured. Monetary damage, 500,000 marks. |
| 13th Aug. | (a) Mannheim (Benz works). Alternative, Speyerdorf (aerodrome). (b) Thionville (railways and workshops). | 104 | 11 (12) | One D.H.9 returned with engine trouble. Formation attacked on return journey, two E.A. were destroyed and one driven down out of control. Three D.H.9's failed to return. The report does not state why the objectives allotted were not attacked. |
| 13/14th Aug. | (a) Buhl aerodrome. Alternative, Volklingen (blast furnaces). (b) Thionville (railways). | 216 | 1 (4) | Three H.P.'s returned owing to bad visibility. They were unable to locate either of the targets allotted. One H.P. bombed Thionville. |
| 14th Aug. | (a) Cologne (explosives works and railways). Alternative, Coblenz (barracks and railways). (b) Offenburg (railways). | 55 | 10 (12) | One D.H.4 returned with engine trouble, one with sick pilot. When over Strassburg large banks of clouds extended as far as could be seen, so leader abandoned the idea of bombing original objective and bombed Offenburg. Formation attacked by E.A. over Strassburg and Offenburg; three of the enemy |

| Date. | Objectives:<br>(a) allotted,<br>(b) actually bombed. | Sqdns. | No. of aeroplanes which bombed. (No. which set out in brackets.) | Remarks. |
|---|---|---|---|---|
| *1918*<br>14th Aug. (*cont.*) | | | | aeroplanes were destroyed and two were driven down out of control. Considerable damage caused to houses. Casualties reported, 2 soldiers wounded. |
| 14/15th Aug. | (*a*) and (*b*) Volkingen (blast furnaces) and Buhl aerodrome. Saarburg (railways). | 216 | 8 (8) | The main attack was made on Buhl aerodrome, but three H.P.'s bombed the alternative target (Volklingen) and Saarburg. |
| 15/16th Aug. | (*a*) and (*b*) Mannheim (Benz, Oppau, and chemical works); Saarbrücken (Burbach works); also Buhl and Boulay aerodromes. | 216 | 8 (10) | One H.P. returned owing to bad visibility, one with engine trouble. Damage caused to workshops. Monetary damage 60,500 marks. Buhl and Boulay aerodromes also attacked. |
| 16th Aug. | (*a*) Cologne (explosive works and railways). Alternatives: (i) Mannheim (Benz works); (ii) Bitche aerodrome. (*b*) Darmstadt. | 55 | 10 (11) | One D.H.4 forced-landed in our lines. Owing to heavy clouds leader decided on Mannheim. On reaching Mannheim, as the wind did not appear to have affected the ground speed of the formation much, the leader decided to carry on and attack Darmstadt. Near Mannheim and Darmstadt D.H.4's attacked by E.A. and three of the enemy aeroplanes were driven down out of control. Three D.H.4's failed to return. Works damaged. Casualties reported, 4 killed, 4 injured. Monetary damage, 88,000 marks. |
| 16/17th Aug. | (*a*) Boulay aerodrome and Mannheim (chemical works). Alternative, Saarbrücken (Burbach works). | 216 | 2 (7) | Five H.P.'s returned with bombs, being unable to locate any target owing to unfavourable weather conditions. |

| Date | Objective | | | Remarks |
|---|---|---|---|---|
| 17/18th Aug. | (b) Boulay aerodrome and Saarburg (railway junction). | 216 | 6 (9) | Two H.P.'s returned owing to bad visibility, one forced-landed in our lines. Casualties reported, 9 killed and many injured. |
| | (a) and (b) Saarbrücken (Burbach works), Forbach (factory) and Black Forest. Boulay aerodrome. | | | |
| 18/19th Aug. | (a) Boulay aerodrome, Cologne (barracks and station), and Frankfurt (barracks and station). Alternatives, Coblenz (barracks and railways) and Mannheim (Benz works). | 216 | 5 (5) | Very adverse weather conditions encountered. Only Boulay aerodrome attacked of the objectives allotted. |
| | (b) Boulay and Buhl aerodromes and Saarbrücken (Burbach works). | | | |
| 19/20th Aug. | (a) and (b) Metz-Sablon (railways). | 97 | 3 (5) | Two H.P.'s returned owing to low clouds. |
| 20th Aug. | (a) and (b) Dillingen. | 99 | 1 (1) | |
| 20/21st Aug. | (a) and (b) Metz-Sablon and Thionville (railways) and Buhl aerodrome. | 97 | 5 (5) | Direct hits obtained on Thionville station which damaged station buildings and tracks, and destroyed four engines. Monetary damage, 12,000 marks. |
| 21/22nd Aug. | (a) and (b) Cologne (railway station), Frankfurt (railway station), Treves (railway junction), and Boulay aerodrome. | 216 | 7 (8) | One H.P. returned with engine trouble. Considerable damage to property in Cologne. Casualties reported, 6 killed and 10 injured. Estimated monetary damage, 477,000 marks. |
| 22nd Aug. | (a) Cologne (barracks and railways). | 55 | 8 (12) | Two D.H.4's returned with engine trouble, one forced-landed, one returned with sick observer. At Treves the leader decided wind too strong for original objective, so proceeded to Coblenz. Casualties reported, 2 killed, 2 injured. |
| | (b) Coblenz (railways and factories), and Wittlick. | | | |
| 22nd Aug. | (a) and (b) Mannheim (chemical works). | 104 | 12 (12) | On the way to the objective and over the objective the D.H.9's were attacked by formations of E.A. During the fighting three |

| Date. | Objectives: (a) allotted, (b) actually bombed. | Sqdns. | No. of aeroplanes which bombed. (No. which set out in brackets.) | Remarks. |
|---|---|---|---|---|
| 1918 22nd Aug.(cont.) | | | | of the enemy aeroplanes were destroyed and one driven out of control. Seven D.H.9's failed to return. |
| 22/23rd Aug. | (a) Ehrang (railway junction) and Volperesweiler aerodrome. (b) Herzing (railway) and Volperesweiler aerodrome. | 97 | 2 (6) | Two H.P.'s returned with engine trouble, two missing. (One shot down in flames.) |
| 22/23rd Aug. | (a) and (b) Saaralben (factories), and Volperesweiler aerodrome. | 100 | 6 (9) | Three F.E.2b's returned with engine trouble. |
| 23rd Aug. | (a) Cologne (barracks and railways. Alternative Coblenz (barracks and railways). (b) Treves (railways). | 55 | 10 (12) | Two D.H.4's returned with engine trouble. Large banks of clouds NE. of Treves prevented the formation reaching the objectives allotted. |
| 23/24th Aug. | (a) and (b) Ehrang (railway junction) and Boulay aerodrome. | 215 | 5 (8) | One H.P. returned with engine trouble, two owing to unfavourable weather conditions. Direct hits obtained on tracks. Only one track could be utilized for 24 hours. |
| 25th Aug. | (a) Cologne (barracks and railways). Alternative, Coblenz (barracks and railways). (b) Luxembourg (railways). | 55 | 6 (6) | When over Treves leader decided wind too strong to reach objectives allotted, so turned and attacked Luxembourg. |
| 25th Aug. | (a) and (b) Bettembourg (railways). | 99 | 12 (12) | |
| 25/26th Aug. | (a) and (b) Mannheim (chemical works) and Boulay aerodrome. | 215 | 5 (6) | One H.P. returned with engine trouble. Chemical works hit and freezing-plant seriously damaged. Monetary damage, 162,500 marks. |

| Date | Target | | | Remarks |
|---|---|---|---|---|
| 25/26th Aug. | (a) Cologne (barracks and station), Frankfurt (railways and works). Boulay aerodrome. (b) Frankfurt (railway station), Boulay aerodrome. | 216 | 2 (8) | Six H.P.'s returned owing to unfavourable weather conditions. The remaining two machines, however, bombed two of the objectives allotted in spite of thunderstorms and heavy rain. Some material damage. |
| 27th Aug. | (a) Bettembourg (railways). (b) Conflans. | 55 | 10 (12) | Two D.H.4's returned with engine trouble. Owing to thick clouds the formations could only reach Conflans. |
| 30th Aug. | (a) Cologne (barracks and railways). Alternative, Coblenz (barracks and railways). (b) Thionville and Conflans. | 55 | 11 (12) | One D.H.4 forced-landed in our lines. Owing to high wind and general unfavourable weather conditions leader decided on Thionville sidings. The formation attacked by E.A. before and after dropping their bombs. In the fighting three of the E.A. were driven down out of control. Four D.H.4's failed to return. |
| 30th Aug. | (a) Bettembourg (railway junction). Alternative, Thionville (railway junction). (b) Conflans (railways) and Doncourt aerodrome. | 99 | 11 (12) | One D.H.9 returned with engine trouble. When near Conflans weather was very cloudy, so one formation attacked Conflans, the other the aerodrome at Doncourt. Formation attacked by E.A. in the vicinity of Conflans and one E.A. was destroyed. |
| 2/3rd Sept. | (a) and (b) Ehrang (railways) and Buhl aerodrome. | 215 | 9 (11) | Two H.P.'s returned with engine trouble. Seven of the H.P.'s bombed Buhl aerodrome. |
| 2/3rd Sept. | (a) and (b) Saarbrücken (Burbach works), and Boulay aerodrome. | 216 | 8 (8) | Direct hits obtained on the Burbach works. One bomb hit the carpenter's shop which was non-effective for fourteen days. Valuable records destroyed. This bomb caused more damage to Burbach works than any other bomb dropped during the War on this target. Monetary damage, 400,000 marks. Casualties reported, 6 killed, 1 injured. Three of the H.P.'s attacked Boulay aerodrome. |
| 3/4th Sept. | (a) and (b) Esch (blast furnaces), Boulay and Morhange aerodromes. | 216 | 7 (7) | One H.P. attacked Esch, the others attacked the aerodromes as detailed. |
| 7th Sept. | (a) Cologne. (b) Ehrang (railway station). | 55 | 6 (6) | |

| Date. | Objectives: (a) allotted, (b) actually bombed. | Sqdns. | No. of aeroplanes which bombed. (No. which set out in brackets.) | Remarks. |
|---|---|---|---|---|
| 1918 7th Sept. | (a) and (b) Mannheim (chemical works). | 99 | 11 (12) | Formation attacked by E.A. over objective and on return journey. One D.H.9 failed to return. |
| 7th Sept. | Ditto. | 104 | 10 (12) | Two D.H.9's returned with engine trouble. One forced-landed in our lines on return journey. This formation was also attacked over the objective and on returning. During the fighting three E.A. were destroyed; three D.H.9's failed to return. |
| 12th Sept. | (a) and (b) Courcelles (railways). | 99 | 2 (2) | Later in the day four D.H.9's attempted another raid on Courcelles but were prevented from reaching the objective by a thunderstorm. |
| 12th Sept. | (a) Courcelles (railways). (b) Orny and Verny (railways). | 99 | 4 (4) | |
| 12th Sept. | (a) Metz–Sablon (railways). (b) Champey. | 104 | 2 (12) | Three D.H.9's returned with engine trouble, one with sick pilot. Very thick clouds encountered and it was decided to abandon raid. Two pilots dropped bombs N. of Champey. Three D.H.9's crashed on our side of the lines on returning. |
| 12/13th Sept. | (a) and (b) Metz–Sablon (railways). | 97 | 2 (2) | Only two H.P.'s set out on this raid. Weather conditions very bad; very strong wind and thick clouds. |
| 12/13th Sept. | Ditto. | 100 | 1 (2) | One of the H.P.'s returned owing to the unfavourable weather conditions. |
| 12/13th Sept. | Ditto. | 215 | 1 (2) | Ditto. |
| 12/13th Sept. | Ditto. | 216 | 4 (8) | One H.P. returned with engine trouble, one with sick observer, two unable to locate objective owing to weather conditions. |
| 13th Sept. | (a) Courcelles (railways). (b) Orny and Verny villages. | 99 | 2 (12) | The target detailed for this day was Courcelles. Many attempts to reach the target were made throughout the day, as detailed |

| Date | Target | Sqn | No. | Remarks |
|---|---|---|---|---|
| | (b) Ars (railways). | 99 | 1 (2) | below, but were unsuccessful owing to very unfavourable weather conditions. At 7.0 a.m. ten out of twelve D.H.9's abandoned the raid, the clouds being too thick. At 12 noon two D.H.9's attempted a raid, one failed to return. |
| | (b) Ars (railways) and Arnaville (M.T. Park). | 99 | 6 (6) | These D.H.9's attacked other railway objectives at 2.30 p.m. |
| | (b) Verny (M.T. Park). | 99 | 2 (2) | Clouds prevented objective being reached at 3.30 p.m. One D.H.9 returned with engine trouble. The formation started out at 4.30 p.m. Owing to thick clouds the leader decided to bomb Arnaville. Over this objective the formation was attacked by E.A. and again on return journey. One E.A. was destroyed and one D.H.9 failed to return. |
| | (b) Arnaville. | 99 | 5 (6) | |
| 13th Sept. | (a) and (b) Metz–Sablon (railways). | 104 | 14 (14) | These D.H.9's carried out individual bombing flights commencing from 12.10 p.m. The clouds were low and very thick, compass courses being steered through the clouds in each case. Railways and villages in vicinity of Metz also attacked. |
| 13/14th Sept. | (a) Mannheim (chemical works), Courcelles and Metz–Sablon (railways). | 215 | 2 (4) | Two H.P.'s returned owing to unfavourable weather. The visibility was extremely bad and all machines encountered thick banks of clouds. |
| | (b) Courcelles (railways). | | | |
| 13/14th Sept. | (a) Boulay and Buhl aerodromes. | 100 | 2 (4) | Two H.P.'s attempted to bomb Boulay aerodrome, but owing to bad visibility and low clouds it could not be reached. Buhl aerodrome was reached but the bomb release would not operate so the aerodrome was machine-gunned. |
| | (b) Courcelles (railways). | | | |
| 14th Sept. | (a) and (b) Ehrang (railways). | 55 | 11 (12) | One D.H.4 forced-landed in our lines. Material damage reported at Treves and neighbourhood. Casualties, 1 killed, 1 injured. |
| 14th Sept. | (a) and (b) Metz–Sablon (railways). | 104 | 10 (12) | Two D.H.9's returned with engine trouble. Shortly after crossing the lines, formation attacked by E.A. One was destroyed and another driven down out of control. |

| Date. | Objectives: (a) allotted, (b) actually bombed. | Sqdns. | No. of aeroplanes which bombed. (No. which set out in brackets.) | Remarks. |
|---|---|---|---|---|
| 1918 14th Sept. (cont.) | | | | Later in the day the same objective was attacked by six D.H.9's of the Squadron. Five had to return with engine trouble. |
| 14th Sept. | (a) and (b) Metz–Sablon (railways). | 104 | 6 (11) | |
| | | 99 | 3 (6) | Two D.H.9's returned with engine trouble. Before reaching objective, formation attacked and one E.A. was driven down out of control. One D.H.9 failed to return. |
| | | | 10 (12) | Later in the day the same objective was attacked by ten D.H.9's of the Squadron; two had to return with engine trouble. |
| 14/15th Sept. | (a) and (b) Metz–Sablon, Kaiserslautern and Courcelles (railways). | 97 | 9 (11) | Two H.P.'s returned with engine trouble. Considerable damage to property at Kaiserslautern and district. Monetary damage, 272,762 marks. Direct hits obtained on Metz–Sablon station; three engines and tracks badly damaged. Casualties reported, 4 killed, 11 injured. |
| 14/15th Sept. | (a) and (b) Metz–Sablon (railways). | 100 | 3 (5) | Two H.P.'s returned with engine trouble. Damage caused to military material in siding, 1 soldier killed. |
| 14/15 Sept. | (a) and (b) Ehrang, Kaiserslautern and Courcelles (railways). | 215 | 9 (13) | Four H.P.'s returned with engine trouble. Two H.P.'s failed to return. |
| 14/15th Sept. | (a) and (b) Saarbrücken, Metz–Sablon and Courcelles (railways), and Frescaty aerodrome. | 216 | 8 (11) | One H.P. returned with engine trouble, one forced-landed in our lines, one failed to return. |
| 15th Sept. | (a) and (b) Stuttgart (Bosche & Daimler works). | 55 | 9 (12) | One D.H.4 returned with engine trouble, two forced-landed in our lines. Formation attacked on return journey and two E.A. were destroyed. Much damage caused to houses. Casualties |

| Date | Target | | | Remarks |
|---|---|---|---|---|
| 15th Sept. | (a) and (b) Metz-Sablon (railways). | 99 | 13 (13) | reported, 12 killed, 11 injured. Monetary damage, 65,000 marks. |
| 15th Sept. | Ditto. | 104 | 12 (12) | When over the objective one of the formations heavily attacked by large number of enemy scouts. In the fighting one E.A. was destroyed and one driven down out of control. Three of the D.H.9's forced-landed in our lines, one crashed on landing and three failed to return. |
| 15/16th Sept. | (a) and (b) Mainz (railway station), Lorquin, Morhange, and Boulay aerodromes. | 97 | 9 (11) | Two H.P.'s returned with engine trouble. |
| 15/16th Sept. | (a) and (b) Karlsruhe (docks and railways), and Buhl aerodrome. | 215 | 5 (6) | One H.P. returned with engine trouble. Damage reported to workshops at Karlsruhe. Casualties, 5 injured. Monetary damage, 60,000 marks. |
| 15/16th Sept. | (a) and (b) Karlsruhe (docks and railways), Metz-Sablon (railways), and Morhange aerodrome. | 216 | 6 (6) | |
| 16th Sept. | (a) and (b) Mannheim (factories and station). | 55 | 5 (6) | One D.H.4 failed to return. |
| 16th Sept. | Ditto. | 110 | 11 (12) | One D.H.9 returned with engine trouble. Formation attacked near objective and one E.A. driven down out of control. Two of the D.H.9's shot down over enemy territory. Some material damage caused. Casualties reported, 1 killed, 8 injured. Monetary damage, 30,000 marks. |
| 16/17th Sept. | (a) and (b) Frankfurt (railways) and Frescaty aerodrome. | 97 | 3 (5) | Two H.P.'s returned with engine trouble. Some material damage was caused at Frankfurt. |
| 16/17th Sept. | (a) and (b) Metz-Sablon (railways). | 115 | 6 (8) | Two H.P.'s returned with engine trouble. One H.P. failed to return. |
| 16/17th Sept. | (a) Frescaty aerodrome, | 215 | .. (5) | Three H.P.'s left to bomb Cologne, one returned with engine |

| Date. | Objectives: (a) allotted, (b) actually bombed. | Sqdns. | No. of aeroplanes which bombed. (No. which set out in brackets.) | Remarks. |
|---|---|---|---|---|
| 1918 16/17 Sept. (cont.) | Cologne (railways). Alternative, Coblenz (railways). (b) See remarks column. | | | trouble, two failed to return; one left to bomb Frescaty but failed to return; one left to bomb Mannheim also failed to return. Reports received from German sources state one bomb fell at Cologne which caused some material damage but no casualties, and that the A.A. fire over the town was heavy; otherwise no further particulars are available. Weather conditions for this night are reported as 'Visibility, fair at first, but thick haze arising later making observation difficult'. |
| 16/17th Sept. | (a) As for 215 Sqdn. (b) Frescaty and Boulay aerodromes, Metz, Treves, Merzig and district (railways). | 216 | 5 (6) | No attempt was made on Cologne or Coblenz. Weather reported as 'Visibility good but becoming misty. Wind strong and gusty'. One H.P. failed to return. A distillery in Treves was hit by a bomb and destroyed and considerable damage was caused to houses. Total casualties reported, 3 killed. Monetary damage 100,000 marks. |
| 20/21st Sept. | (a) and (b) Karlsruhe (gas works), Boulay and Buhl aerodromes. | 97 | 9 (9) | Works damaged. Monetary damage, 50,000 marks. |
| 20/21st Sept. | (a) and (b) Mannheim (Lanz works), Karlsruhe (gas works and bridges), Saarbrücken (Burbach works). | 100 | 3 (4) | One H.P. returned with engine trouble. |
| 20/21st Sept. | (a) and (b) Mannheim (factories) and Frescaty aerodrome. | 216 | 7 (9) | Two H.P.'s returned with engine trouble. (*Note*: Casualties reported in Mannheim-Karlsruhe district on this night, 1 killed, 3 injured.) |

| Date | Target | | Remarks |
|---|---|---|---|
| 21/22nd Sept. | (a) and (b) Saarbrücken (Burbach works). Alternative, Morhange (aerodrome) and Leiningen (railway station). | 115    3 (4) | One H.P. returned with engine trouble. In addition to the alternative target attacked (Morhange), some 14 112-lb. bombs dropped on Leiningen station by a H.P. which missed the objective owing to clouds. |
| 21/22nd Sept. | (a) and (b) Hagendingen (blast furnaces). | 215    3 (3) | |
| 21/22nd Sept. | (a) and (b) Rombach (factories and railways). | 216    6 (6) | |
| 25th Sept. | (a) and (b) Kaiserslautern (munition factories). | 55    12 (12) | Formation attacked on return journey, two E.A. destroyed and two driven down out of control. Three D.H.4's failed to return. Heavy A.A. fire encountered. Some damage caused to houses, but no casualties. |
| 25th Sept. | (a) and (b) Frankfurt (works and railways). | 110    11 (12) | One D.H.9a forced-landed. Formation attacked practically the whole way out and home, and two E.A. driven down out of control. Four of the D.H.9a's failed to return. Considerable material damage caused to workshops. Casualties reported, 1 killed, 7 injured. Monetary damage, 146,000 marks. |
| 26th Sept. | (a) and (b) Adun-le-Roman (railways). | 55    6 (12) | Four D.H.4's returned with engine trouble, two forced-landed in our lines. |
| 26th Sept. | (a) Thionville (railways and bridges). (b) Metz-Sablon (railways). | 99    6 (10) | Formation left to bomb Thionville, but two D.H.9's returned with engine trouble, and one lost formation. On the outward journey the D.H.9's were attacked by large enemy formations estimated between 30–40. Two were destroyed and two driven down over enemy territory. The remaining six machines bombed Metz. Of these four failed to return, making a total of five D.H.9's missing during this raid. |
| 26th Sept. | (a) and (b) Metz-Sablon (railways). | 104    10 (12) | Two D.H.9's returned with engine trouble. Over the objective the formation was attacked by E.A. and one of the D.H.9's was shot down. |

| Date. | Objectives: (a) allotted, (b) actually bombed. | Sqdns. | No. of aeroplanes which bombed. (No. which set out in brackets.) | Remarks. |
|---|---|---|---|---|
| 1918 26/27th Sept. | (a) and (b) Metz–Sablon, Mezières and Ars (railways). | 97 | 5 (6) | One H.P. returned with engine trouble. |
| 26/27th Sept. | (a) and (b) Metz–Sablon and Thionville (railways). | 100 | 2 (2) | |
| 26/27th Sept. | (a) and (b) Metz–Sablon and Thionville (railways). Plappeville. | 115 | 2 (4) | One H.P. returned with engine trouble, one owing to unfavourable weather conditions. |
| 26/27th Sept. | (a) and (b) Metz–Sablon (railways) and Frescaty aerodrome. | 215 | 3 (6) | One H.P. returned with engine trouble, two owing to the unfavourable weather. |
| 26/27th Sept. | (a) and (b) Mezières and Metz–Sablon (railways). | 216 | 4 (8) | Four H.P.'s returned owing to unfavourable weather conditions. (*Note*: During the raids on the railways at Metz–Sablon on the night of the 26/27th the Clemens bridge was damaged and many bombs fell in station area considerably dislocating traffic and causing stoppages varying between 12–24 hours. Casualties reported, 3 killed, 6 injured.) |
| 30th Sept./1st Oct. | (a) and (b) Mezières (railways). | 97 | 2 (8) | Six H.P.'s returned owing to unfavourable weather. Very dark, low clouds. Visibility exceedingly bad. |
| 30th Sept./1st Oct. | (a) and (b) Saarbrücken (Burbach works). | 100 | 1 (3) | Two H.P.'s returned owing to bad visibility. |
| 30th Sept./1st Oct. | (a) Audun, Mezières, and Thionville (railways). (b) Abandoned. | 115 | .. (6) | All forced to return owing to weather conditions. |
| 30th Sept./1st Oct. | (a) and (b) Metz–Sablon (railways) and Frescaty aerodrome. | 215 | 2 (5) | Three H.P.'s returned owing to unfavourable weather conditions. |

| Date | Target | | | Remarks |
|---|---|---|---|---|
| 30th Sept./1st Oct. | (a) Mezières (railways). (b) Abandoned. | 216 | .. (4) | All forced to return owing to weather conditions, one crashed on landing. |
| 1st Oct. | (a) Cologne (b) Treves and Luxembourg. | 110 | 6 (12) | Two D.H.9a's returned with engine trouble, four owing to unfavourable weather conditions. Bursts not observed. (*Note*: 55 and 104 Squadrons attempted raids which had to be abandoned owing to weather conditions.) |
| 3/4th Oct. | (a) Mezières (railways). Alternative, Thionville. (b) Metz–Sablon. | 97 | 1 (4) | Three H.P.'s returned owing to bad visibility, remaining H.P. bombed Metz, bursts not observed. |
| 3/4th Oct. | (a) and (b) Metz–Sablon (railways). Morhange aerodrome. | 100 | 2 (5) | Three H.P.'s returned owing to unfavourable weather conditions. |
| 3/4th Oct. | (a) and (b) Metz–Sablon (railways). | 115 | 1 (5) | Two H.P.'s returned with engine trouble, one owing to unfavourable weather conditions, and one lost direction. |
| 3/4th Oct. | Ditto. | 216 | 2 (7) | Five H.P.'s returned owing to unfavourable weather conditions. |
| 5th Oct. | Ditto. | 104 | 12 (12) | Metz–Sablon station hit, rolling-stock damaged and cellar used as an air-raid shelter hit and destroyed. Casualties reported, 12 killed, 23 injured. |
| 5th Oct. | (a) Cologne (factories). Alternatives, Coblenz (bridges) or Ehrang (railways). (b) Kaiserslautern and Pirmasens. | 110 | 8 (12) | Owing to unfavourable weather and heavy fighting, the targets allotted could not be reached. The formation was continually attacked on the outward and return journey; one E.A. was destroyed and three driven down out of control. Four D.H.9a's failed to return. Bursts unobserved. |
| 5/6th Oct. | (a) Mezières or Thionville (railways). (b) Metz–Sablon and Courcelles (railways). | 97 | 5 (7) | One H.P. returned damaged, one owing to unfavourable weather conditions. Unable to reach objectives owing to bad visibility. |
| 5/6th Oct. | (a) and (b) Mezières (railways), Saarbrücken (Burbach works), Morhange aerodrome. | 100 | 4 (4) | |
| 5/6th Oct. | (a) and (b) Thionville and Metz–Sablon (railways). | 115 | 2 (5) | Two H.P.'s returned with engine trouble, one owing to unfavourable weather conditions. |

| Date. | Objectives: (a) allotted, (b) actually bombed. | Sqdns. | No. of aeroplanes which bombed. (No. which set out in brackets.) | Remarks. |
|---|---|---|---|---|
| 1918 5/6th Oct. | (a) Mezières or Audun (railways). | 216 | 5 (7) | Two H.P.'s returned owing to bad visibility. |
| 9th Oct. | (b) Metz–Sablon (railways). (a) and (b) Metz–Sablon (railways). | 99 | 11 (12) | One D.H.9 returned with engine trouble. |
| 9th Oct. 9/10th Oct. | Ditto. (a) Mezières or Thionville (railways). | 104 97 | 13 (13) 3 (7) | Two H.P.'s returned owing to bad visibility, one crashed on aerodrome, one forced-landed in our lines. Visibility extremely bad. |
| 9/10th Oct. | (b) Metz–Sablon (railways). (a) Saarbrücken (Burbach works), Mezières (railways). (b) Mezières (railways). | 100 | 2 (5) | Three H.P.'s detailed for Saarbrücken returned owing to unfavourable weather conditions. |
| 9/10th Oct. | (a) and (b) Thionville (railways) and Morhange aerodrome. | 115 | 5 (6) | One H.P. returned owing to the weather. |
| 9/10th Oct. | (a) and (b) Metz–Sablon (railways), and Frescaty aerodrome. | 215 | 2 (4) | One H.P. returned with engine trouble, one owing to the weather. |
| 9/10th Oct. | (a) Mezières or Audun (railways). (b) Metz–Sablon, Mezières, and Thionville (railways). | 216 | 6 (8) | Two H.P.'s returned owing to weather conditions. Owing to very bad visibility bursts unobserved in most cases. (*Note*: In the raids on Metz one bomb fell on the powder magazine in the Metz Wiese island. A large explosion occurred and a fire was burning for four days. Damage estimated at 1,000,000 marks.) |

| Date | Target | | | Remarks |
|---|---|---|---|---|
| 10th Oct. | (a) and (b) Metz–Sablon (railways). | 99 | 8 (12) | Four D.H.9's returned with engine trouble. |
| 10th Oct. | Ditto. | 104 | 12 (12) | One returned owing to bad weather, crashed on landing. |
| 10/11th Oct. | (a) and (b) Thionville (railways). | 100 | 1 (2) | All returned owing to unfavourable weather conditions. |
| 10/11th Oct. | (a) Longuyon and Metz–Sablon (railways). (b) Abandoned. | 115 | .. (4) | |
| 10/11th Oct. | (a) and (b) Mezières (railways) and Rombach. | 216 | 3 (8) | Two H.P.'s returned owing to bad visibility, two with engine trouble, one forced-landed in our lines. |
| 18th Oct. | (a) and (b) Metz–Sablon (railways). | 99 | 8 (12) | Three D.H.9's returned with engine trouble, one lost formation and returned. |
| 18/19th Oct. | (a) Metz–Sablon and Saarbrücken. (b) Raid abandoned. | 97 | .. (2) | H.P.'s returned owing to low clouds and fog. |
| 18/19th Oct. | (a) Kaiserslautern (railways and factories), Essen (Krupp works) and Mezières (railways). (b) Saarburg. | 100 | 1 (2) | Only two H.P.'s attempted to raid, one returned owing to the very bad visibility. |
| 18/19th Oct. | (a) Metz–Sablon and Saarbrücken. | 115 | .. (2) | H.P.'s returned owing to the weather. |
| 18/19th Oct. | Ditto. | 215 | .. (1) | Thick clouds and haze made bombing impossible. |
| 18/19th Oct. | (a) Cologne. | 216 | .. (2) | Ditto. |
| 21st Oct. | (b) Thionville (railways). | 55 | 7 (12) | Four D.H.4's returned with engine trouble, one forced-landed in our lines. Owing to banks of cloud and thick mist, leader decided to bomb Thionville—observation practically impossible. |
| 21st Oct. | (a) and (b) Metz–Sablon (railways). | 99 | 11 (11) | |
| 21st Oct. | Ditto. | 104 | 14 (14) | |
| 21st Oct. | (a) Cologne (factories). Alter- | 100 | 5 (13) | One D.H.9a returned with engine trouble. Owing to heavy |

| Date. | Objectives: (a) allotted, (b) actually bombed. | Sqdns. | No. of aeroplanes which bombed. (No. which set out in brackets.) | Remarks. |
|---|---|---|---|---|
| 1918 21st Oct. (cont.) | natives, Coblenz (bridges) or Ehrang (railways). (b) Frankfurt (factories and railways near). | | | clouds formation broken up. Some E.A. were encountered but were driven off. Seven D.H.9a's failed to return. (German reports account for five as brought down over enemy territory N. of Frankfurt as follows: one in flames at Dillenburg, one at Berenbach, one near Lauterbach, two near Weiler.) |
| 21/22nd Oct. | (a) and (b) Kaiserslautern (railways) and district. | 97 | 3 (4) | One H.P. returned with engine trouble. One munition factory (Greist) was completely destroyed by a 1,650-lb. bomb. Monetary damage to factory estimated at 500,000 marks. Much private property destroyed in Kaiserslautern. Casualties reported, 1 killed, 1 injured. Total estimated monetary damage, 671,000 marks. |
| 21/22nd Oct. | (a) and (b) Mezieres and Kaiserslautern (railways). | 100 | 3 (6) | One H.P. returned with engine trouble, two owing to unfavourable weather conditions. |
| 23rd Oct. | (a) Cologne (factories near). Alternatives, Coblenz (bridges) or Ehrang (railways). (b) Metz-Sablon (railways). | 55 | 11 (12) | One D.H.4 lost formation. When over Metz leader experienced engine trouble so bombs were released on the railways. |
| 23rd Oct. | (a) and (b) Metz-Sablon. | 99 | 11 (12) | One D.H.9 returned with engine trouble. Formation attacked over objective and one E.A. driven down out of control. |
| 23rd Oct. | Ditto. | 104 | 11 (12) | One D.H.9 returned with engine trouble. This formation also attacked by E.A. over objective. One E.A. was destroyed and two driven down out of control. One D.H.9 missing (Note: During the raids on the 23rd much damage was caused to rail- |

| Date | Target | | | Remarks |
|---|---|---|---|---|
| 23/24th Oct. | (a) Mannheim (chemical works), Essen (Krupps works), Mezières (railways), Frankfurt (factories). Alternatives, Kaiserslautern and Saarbrücken. (b) Saarbrücken (Burbach works), and Metz. | 100 | 4 (6) | way tracks and much material damage caused in the Metz area). Two H.P.'s returned owing to bad weather. Mannheim and district targets could not be reached owing to unfavourable weather conditions. |
| 23/24th Oct. | (a) and (b) Saarbrücken (railways) and Kaiserslautern (station). | 215 | 4 (6) | Two H.P.'s returned with engine trouble. |
| 23/24th Oct. | (a) and (b) Mannheim (chemical works), Coblenz, Saarbrücken (factories and railways) and Metz–Sablon (railways). | 216 | 5 (9) | Three H.P.'s returned with engine trouble, one owing to unfavourable weather conditions. Damage caused to houses and workshops. Casualties reported, one killed. Monetary damage, 44,000 marks. |
| 23/24th Oct. | (a) Mannheim (chemical works), Essen (Krupp works), Cologne (railway bridge over Rhine at), Saarbrücken or Kaiserslautern (railways). (b) Wiesbaden, and aerodrome near Saarbrücken. | 97 | 1 (6) | Five H.P.'s returned. Visibility very bad. One 1,650-lb. bomb which dropped in centre of town of Wiesbaden caused very considerable damage to property. Casualties reported, 13 killed and 36 injured. |
| 28/29th Oct. | (a) Ecouviez (railway junction), Cologne (bridge), Liège (Val Benoit bridge). (b) Treves, Thionville, and Saarbrücken. | 216 | 6 (8) | Two H.P.'s returned with engine trouble. Owing to the bad visibility many of the bursts unobserved. |
| 28/29th Oct. | (a) and (b) Mannheim (chemical works). | 97 | 1 (5) | Four H.P.'s returned owing to unfavourable weather conditions. One H.P. carried on and dropped a 1,650-lb. bomb on the target; it was wrecked, however, on the return journey, pilot and observer being killed. |

| Date. | Objectives: (a) allotted, (b) actually bombed. | Sqdns. | No. of aeroplanes which bombed. (No. which set out in brackets.) | Remarks. |
|---|---|---|---|---|
| *1918* | | | | |
| 28/29th Oct. | (a) and (b) Longuyon (railways). | 100 | 1 (7) | Only one H.P. succeeded in reaching the objective. The others detailed for Mannheim and Frankfurt returned owing to unfavourable weather conditions. |
| 28/29th Oct. | (a) and (b) Longuyon and Thionville (railways). | 115 | 3 (4) | One H.P. returned owing to weather conditions. |
| 28/29th Oct. | (a) Ecouviez (railways), Cologne (bridge). (b) Ecouviez (railways). | 215 | 1 (9) | Only one H.P. bombed Ecouviez. Five returned owing to bad weather, heavy ground mist and fog, and three with engine trouble. |
| 29th Oct. | (a) and (b) Longuyon (railways). | 55 | 10 (12) | Two D.H.4's returned with engine trouble. Formation attacked by E.A. and one E.A. driven down out of control. |
| 29th Oct. | Ditto. | 99 | 12 (12) | |
| 29/30th Oct. | (a) and (b) Mannheim (chemical works) and Hagenau aerodrome. | 97 | 2 (2) | |
| 29/30th Oct. | (a) Mannheim (chemical works), Saarbrücken (Burbach works). (b) Offenburg. | 100 | 2 (5) | Two H.P.'s returned owing to unfavourable weather, one with engine trouble. Visibility rendered observation of bursts impossible. |
| 29/30th Oct. | (a) Mannheim and Saarbrücken. (b) Abandoned. | 115 | .. (5) | One H.P. returned with engine trouble, four owing to unfavourable weather. |
| 29/30th Oct. | (a) and (b) Mannheim (chemical works), Thionville and Offenburg (railways). | 215 | 4 (8) | One H.P. returned with engine trouble, three owing to bad weather. Very heavy A.A. fire encountered. One H.P. missing. |
| 29/30th Oct. | (a) and (b) Mannheim (chemical | 216 | 6 (8) | One H.P. returned with engine trouble, one owing to bad |

| Date | Target | | Remarks |
|---|---|---|---|
| | works), Saarbrücken (Burbach works). | | weather. (*Note:* Considerable damage was caused to the workshops in the Mannheim district during the raids on this night. Monetary damage, 178,000 marks. German reports state that on the homeward journey from Mannheim and neighbouring objectives the towns of Heidelberg, Hockenheim, Durkheim and others in this area were bombed and considerable damage caused to houses and a university building in Heidelberg. Total casualties reported, 5 killed, 30 injured. Pirmasens, it is stated, was also attacked, when 5 people were killed and injured.) |
| 30/31st Oct. | (*a*) and (*b*) Karlsruhe (railway works), Saarbrücken (Burbach works). | 100   2 (3) | One H.P. returned with engine trouble. Some material damage reported at Pirmasens. Casualties, 3 killed, 9 injured. |
| 30/31st Oct. | (*a*) Gaggenau (aero works), Karlsruhe (railway works), Saarbrücken (Burbach works). (*b*) Baden, also Morhange and Hagenau aerodromes. | 115   3 (3) | Visibility very poor. |
| 31st Oct. | (*a*) Cologne (Deutz works). (*b*) Bonn (railways), Treves (railways), Frescaty aerodrome. | 55   8 (12) | Three D.H.4's returned with engine trouble, one lost formation. Owing to weather conditions, low clouds and poor visibility, the leader decided to turn at Bonn and attack the railways. Nine bombs fell on the targets at Bonn; 29 were killed (16 by one bomb), and 57 wounded. |
| 2nd Nov. | (*a*) and (*b*) Avricourt (railways). | 99   1 (1) | This D.H.9a left for the purpose of a cloud bomb raid. Three 112-lb. bombs were dropped on Avricourt junction and on a small dump to the west of the junction. |
| 3rd Nov. | (*a*) Cologne, Coblenz, or Ehrang. (*b*) Saarburg (railways). | 55   9 (12) | Three D.H.4's returned with engine trouble. Owing to the clouds it was impossible to reach objectives allotted. |
| 3rd Nov. | (*a*) Buhl aerodrome. (*b*) Lorquin (railways). | 104   9 (12) | Two D.H.9's returned with engine trouble, one with sick pilot. Owing to the clouds it was impossible to locate the aerodrome at Buhl. |

| Date. | Objectives: (a) alloted, (b) actually bombed. | Sqdns. | No. of aeroplanes which bombed. (No. which set out in brackets.) | Remarks. |
|---|---|---|---|---|
| 1918 5/6th Nov. | (a) Morhange and Frescaty aerodromes. (b) Dieuze (camp near) and Frescaty aerodrome. | 115 | 2 (2) | Owing to engine trouble one H.P. bombed Dieuze. |
| 6th Nov. | (a) and (b) Saarbrücken (Burbach works) and Hattigny aerodrome. | 55 | 6 (6) | Formation attacked by E.A. over objective. One E.A. was destroyed and two driven down out of control. One D.H.4 failed to return. |
| 9th Nov. | (a) and (b) Bensdorf (railways). | 55 | 1 (6) | Five D.H.4's returned owing to unfavourable weather conditions, low clouds and mist. |
| 9th Nov. | (a) and (b) Chateau Salins (motor transport, &c.). | 99 | 2 (3) | One D.H.9a returned owing to the weather. |
| 9th Nov. | (a) and (b) Lorquin and Rachicourt (railways). | 104 | 2 (5) | Two D.H.9's forced-landed in our lines. One returned owing to adverse weather. |
| 10th Nov. | (a) Cologne (Deutz works). (b) Ehrang (railways). | 55 | 11 (12) | One D.H.4 returned having missed objective. Low clouds, mist, and heavy A.A. fire prevented formation reaching allotted objective. One D.H.4 failed to return. |
| 10/11th Nov. | (a) and (b) Metz–Sablon (railways) and Frescaty aerodrome. | 216 | 2 (3) | One H.P. returned owing to bad visibility. At Metz 5 people were killed and 7 injured by bombs dropped on main street. |

## APPENDIX XIV

### VOLKLINGEN STEEL WORKS. ANALYSIS OF DAMAGE CAUSED BY AIR RAIDS IN 1916, 1917, AND 1918

| Date. | | No. of alarms. | Due to wages.* | Deficit in tons. | Losses due to rise in wages and fall of output. | | | | | Other deficits. | | | Cost of repairs (in Marks). | |
|---|---|---|---|---|---|---|---|---|---|---|---|---|---|---|
| Year. | Month. | | Marks. | Tons. | 21 cm. shells. | 7.7 and 10 cm. shells. | T.M. shells. | Shell bases. | Shell beads. | Steel helmets. | Body armour. | Gun carriages. | Estimate. | Actual. |
| 1916 | September | 7 | 11,356 | 1,016 | 144 | 157 | 35 | 75 | 80 | .. | .. | .. | .. | 42,171 |
| ,, | November | 14 | 16,461 | 1,713 | 150 | 237 | 678 | 560 | 330 | .. | .. | .. | 56,189 | 5,575 |
| ,, | December | 4 | 3,448 | 559 | 60 | 170 | 190 | 40 | .. | .. | .. | .. | 4,670 | 81 |
| 1917 | January | 4 | 2,568 | 326 | 54 | 105 | 131 | .. | 50 | .. | .. | .. | 93 | .. |
| ,, | February | 9 | 7,682 | 1,522 | 102 | 270 | 182 | 125 | 45 | .. | .. | .. | 764,572 | 1,015,174 |
| ,, | March | 7 | 47,278 | 1,310 | 98 | 220 | .. | .. | 40 | .. | .. | .. | .. | .. |
| ,, | April | 1 | 58,592 | 82 | .. | .. | .. | .. | .. | .. | .. | .. | .. | .. |
| ,, | May | 1 | 9,462 | 2,952 | 34 | .. | 80 | .. | .. | 180 | .. | .. | 82,982 | 777,737 |
| ,, | June | 2 | 6,335 | 739 | 80 | 40 | .. | .. | 400 | .. | 220 | .. | 450 | 868 |
| ,, | July | 3 | 507 | 79 | 13 | 20 | .. | .. | 60 | .. | 20 | .. | .. | .. |
| ,, | August | 2 | 94 | .. | .. | .. | .. | .. | .. | .. | .. | .. | .. | .. |
| ,, | September | 7 | 1,216 | 351 | 16 | .. | .. | .. | .. | .. | .. | 180 | 57,330 | 71,865 |
| ,, | October | 17 | 25,609 | 4,068 | 202 | .. | .. | .. | .. | .. | .. | .. | .. | .. |
| ,, | November | .. | .. | .. | .. | .. | .. | .. | .. | .. | .. | .. | .. | .. |
| ,, | December | 4 | 1,770 | 160 | 26 | .. | .. | .. | .. | .. | .. | 35 | .. | .. |
| 1918 | January | 11 | 10,418 | 978 | 105 | .. | .. | .. | .. | 20 | .. | 360 | .. | .. |
| ,, | February | 11 | 9,920 | 958 | 84 | .. | .. | .. | .. | .. | .. | 316 | 164 | 1,055 |
| ,, | March | 8 | 4,765 | 389 | 44 | .. | .. | .. | .. | .. | .. | 170 | .. | .. |
| ,, | April | 1 | .. | .. | .. | .. | .. | .. | .. | .. | .. | .. | .. | .. |

* This loss represents increased wages which had to be paid to compensate workers for air-raid alarms.

| | | | Losses due to rise in wages and fall of output. | | | | | | | Other deficits. | | | Cost of repairs (in Marks). | |
|---|---|---|---|---|---|---|---|---|---|---|---|---|---|---|
| Date. | | No. of alarms. | Due to wages.* | Deficit in tons. | 21 cm. shells. | 7·7 and 10 cm. shells. | T.M. shells. | Shell bases. | Shell heads. | Steel helmets. | Body armour. | Gun carriages. | | |
| Year. | Month. | | Marks. | Tons. | | | | | | | | | Estimate. | Actual. |
| 1918 | May | 13 | 15,251 | 1,651 | 103 | .. | .. | .. | .. | .. | .. | 330 | 36,240 | ? |
| ,, | June | 33 | 21,045 | 2,050 | 179 | 140 | .. | 140 | .. | .. | .. | 575 | .. | .. |
| ,, | July | 47 | 29,960 | 2,106 | 143 | .. | .. | .. | .. | .. | .. | 586 | .. | .. |
| ,, | August | 50 | 29,981 | 2,881 | 229 | .. | .. | .. | .. | .. | .. | 780 | 420 | ? |
| ,, | September | 44 | 26,299 | 2,622 | 189 | .. | .. | .. | .. | .. | .. | 710 | 23,798 | ? |
| ,, | October | 23 | 15,660 | 1,768 | 128 | .. | .. | .. | .. | .. | .. | 425 | .. | .. |
| ,, | November | 4 | 3,847 | 400 | 32 | .. | .. | .. | .. | .. | .. | 30 | ? | ? |
| | Totals | 327 | 350,524 | 30,680 | 2,215 | 1,364 | 1,296 | 940 | 1,005 | 200 | 240 | 4,497 | 1,026,909 (incomplete) | ? |

* This loss represents increased wages which had to be paid to compensate workers for the air-raid alarms.

# APPENDIX XV

## STATE OF INDEPENDENT FORCE, ROYAL AIR FORCE

(formed with effect from 6th June 1918).
Major-General Sir H. M. Trenchard.

### VIII Brigade*
(formed with effect from 1st Feb. 1918).
Brigadier-General C. L. N. Newall.

---

### 41st Wing*
(formed with effect from 11th Oct. 1917).
Lieut.-Col. C. L. N. Newall.
Lieut.-Col. J. E. A. Baldwin (28.12.17).
Lieut.-Col. L. A. Pattinson (22.9.18).

| No. of Sqdn. | Date of arrival in I.F. Area. | Commanding Officer. |
|---|---|---|
| 55 (D.H.4.) | 11.10.17. | Major J. E. A. Baldwin. Major A. Gray (7.1.18). Major B. J. Silly (20.9.18). Major L. A. Pattinson. |
| 99 (D.H.9.) | 3.5.18. | Capt. P. E. Welchman (23.9.18). Capt. W. D. Thom (28.9.18). Major C. R. Cox (5.11.18). |
| 104 (D.H.9.) | 20.5.18. | Major J. C. Quinnell. |

Total aircraft on charge 59.
11.11.18.

---

### 83rd Wing
(formed with effect from 1st July 1918).
Lieut.-Col. J. H. A. Landon.

| No. of Sqdn. | Date of arrival in I.F. Area. | Commanding Officer. |
|---|---|---|
| 97 (H.P.). | 9.8.18. | Major V. A. Albrecht. |
| 100 (H.P.). | 11.10.17 | Major M. G. Christie. Major W. J. Tempest (11.12.17). Major C. G. Burge (16.6.18). |
| 115 (H.P.). | 31.8.18. | Major W. E. Gardiner. |
| 215 (H.P.). | 19.8.18 | Major J. F. Jones. |
| 216 (formerly 'A' Sqdn., R.N.A.S. later 16 (Naval) Sqdn).(H.P.). | 17.10.17. | Sqdn.-Cdr. K. S. Savory. Sqdn.-Cdr. H. A. Buss (28.1.18). Major W. R. Read (1.9.18). |

Total aircraft on charge 49.
11.11.18.

---

### 88th Wing
(formed with effect from 17th Oct. 1918).
Lieut.-Col. W. D. Beatty.

| No. of Sqdn. | Date of arrival in I.F. Area. | Commanding Officer. |
|---|---|---|
| 45 (Sopwith 'Camel'). | 22.9.18. | Major J. A. Crook. Major A. M. Miller (21.10.18). |
| 110 (D.H.9). | 31.8.18. | Major H. R. Nicholl. |

Total aircraft on charge 32.
11.11.18.

---

* On the 1st of February 1918 the bombing force (41st Wing) was raised to the status of a Brigade under Brigadier-General C. L. N. Newall. This Brigade, although formed on the 28th of December 1917, did not actually take over until 1st February 1918.

## APPENDIX XVIII

### COMPARISON OF ANTI-SUBMARINE FLYING OPERATIONS BETWEEN GROUPS NOS. 9, 10, AND 18
### FROM 1st JULY 1918 TO 30th SEPTEMBER 1918

| | Group No. 9 (Plymouth). | | | Group No. 10 (Portsmouth). | | | Group No. 18 (East Coast of England). | | | Grand total. | | |
|---|---|---|---|---|---|---|---|---|---|---|---|---|
| | Sea-planes. | Aero-planes. | Air-ships. | Sea-planes. | Aero-planes. | Air-ships. | Sea-planes. | Aero-planes. | Air-ships. | Sea-planes. | Aero-planes. | Air-ships. |
| Total no. of machines available for War Service | 2,311 | 3,445 | 619 | 2,063 | 1,448 | 615 | 1,781 | 4,256 | 745 | 6,155 | 9,149 | 1,979 |
| Daily average no. of War Service machines available | 25 | 37 | 6·7 | 22 | 16 | 7·1 | 19 | 46 | 8·1 | 66 | 99 | 22 |
| Total no. of days operated | 84 | 89 | 67 | 87 | 87 | 65 | 90 | 91 | 35 | 261 | 267 | 167 |
| Total no. of hours operated | 2,636 | 1,939 | 2,818 | 4,499 | 2,060 | 2,598 | 2,107 | 4,447 | 1,205 | 9,242 | 8,446 | 6,621 |
| Average duration of flight per machine available | H. M. 1 8 | Min. 33 | H. M. 4 32 | H. M. 2 10 | H. M. 1 23 | H. M. 3 39 | H. M. 1 11 | H. M. 1 27 | H. M. 1 37 | H. M. 1 30 | Min. 55 | H. M. 3 16 |
| No. of attacks of merchant shipping by enemy submarines when aircraft were operating | 42 | 46 | 31 | 15 | 16 | 12 | 67 | 69 | 31 | 124 | 131 | 74 |
| No. of sightings of enemy submarines by aircraft | 7 | 7 | Nil | 7 | 2 | 4 | 21 | 32 | 5 | 35 | 41 | 9 |
| No. of attacks on enemy submarines by aircraft | 5 | 6 | Nil | 6 | Nil | 2 | 9 | 16 | 4 | 20 | 22 | 6 |
| No. of days not operated by aircraft | 8 | 3 | 25 | 5 | 5 | 27 | 2 | 1 | 57 | 15 | 9 | 109 |
| No. of attacks on merchant shipping by enemy submarines when aircraft were not operating | 4 | Nil | 15 | 1 | Nil | 4 | 2 | Nil | 38 | 7 | Nil | 57 |
| Hours of flight per enemy submarine sighted | 376 | 277 | .. | 642 | 1,030 | 649 | 100 | 138 | 241 | 264 | 206 | 735 |

# APPENDIX XIX

## A SHORT REVIEW OF THE SITUATION IN THE AIR ON THE WESTERN FRONT AND A CONSIDERATION OF THE PART TO BE PLAYED BY THE AMERICAN AVIATION

*Memorandum of H.Q.R.F.C., France, December 1917*

1. To appreciate the position as it is to-day it is necessary to go back to the commencement of the SOMME battle on 1st July 1916. There is no doubt whatever that we caught the German aviation at a disadvantage at that time and that it had been neglected and allowed to stagnate and that there was little driving power in the superior command.

We realized at once that this state of affairs would not be allowed to last and based our demands for the expansion of the R.F.C. in France on the certainty that the enemy would put forward his best efforts to retrieve the position.

This he did almost at once by placing Lieut.-General von Hoeppner in command of his aviation and giving him undoubtedly wide powers to reorganize and develop it and at the same time mobilizing his resources to carry out the necessary expansion. The result was apparent even in the autumn of 1916 and has been increasingly evident throughout the present year.

Meanwhile our expansion has not been as rapid as was hoped, and we are still far short of the programme put forward in the summer and autumn of 1916.

The result is that our ascendancy has become relatively less and less, and to-day, although in some directions we still continue to do more work, we cannot any longer, as regards the fighting which makes such work possible, claim superiority over the enemy.

The very success which we achieved last year has enabled the enemy, by copying our methods, to reach equal efficiency in a shorter period of time and to pass from the purely defensive attitude he adopted on the SOMME to the offensive tactics which are the only road to success in the air.

There can be no doubt that the enemy realizes to the full the supreme importance of aviation and the value of the progress he has made and that he will use every endeavour, not only to maintain that progress but to increase his advantage, with a view to gaining the ascendancy over us.

2. There is one way and one way only to defeat his endeavours and that is to pursue our offensive policy with the utmost determination and vigour. He will undoubtedly do the same and the result will depend upon which has the means and the nerve to maintain the pressure longest. If we allow ourselves, in view of his increased activity with bomb and machine gun, to be persuaded to adopt a defensive attitude, we shall simply be paving the way to certain defeat. The whole experience of the war—French, German, and British—proves this without the slightest possibility of doubt.

3. To maintain a vigorous offensive next year three requirements must be fulfilled:
  (i) a sufficiency of personnel,
  (ii) efficient methods of training and employment,
  (iii) a sufficiency of machines of suitable types.

As regards the first, there is, I understand, no cause for anxiety; as regards the second, the exertions of the present D.G.M.A., when in command of the Training Division, have resulted in a considerably higher standard of training throughout the R.F.C. Experienced officers are continually being sent home from the front to ensure that methods and teaching are based on the most recent experience, and the officer now commanding the Training Division and his two Senior Staff Officers have only lately relinquished commands on the battle front. Everything possible is therefore being done to ensure efficient and up-to-date instruction. All this will be of little value, however, unless the Training Division is allotted and kept supplied with a sufficiency of machines, and in our programme of future development it is essential to bear this in mind.

As regards the supply of machines to France, the C.-in-C. has lately put forward a programme of his estimated requirements up to the summer of 1919, but it must be remembered that the struggle will be at its height at the earliest moment the weather permits next spring and that any leeway then lost will be difficult, if not impossible, to make up subsequently. If the enemy once succeeds in putting us in a position of inferiority our moral effect will suffer, our casualties will increase, and as a consequence of the latter our programme of development will be upset.

It is imperative, therefore, to concentrate all efforts on being as strong as possible next spring. Flying during the winter will be reduced to a minimum necessary for training and operations, and every endeavour will be made to husband our existing resources. Orders have been issued that every squadron will repair machines to the utmost of its ability so that depots may be freed to build up the reserve.

Three points are of great importance as affecting our fighting strength next spring:
  (i) All replacements must be done during the winter. Re-equipping a squadron puts it out of action for a month and it is better to continue with an inferior machine than to attempt to replace once active operations have commenced.
  (ii) A new squadron which can have a month's experience overseas before being actively employed is of far greater value and experiences far less casualties than a squadron which arrives and has to be put straight into the fight during active operations.
  (iii) The search for perfection defeats its own ends. We must be sufficiently strong minded, once we have decided on a design, to refuse to consider minor improvements which delay output without improving the machines to any appreciable extent. A machine of good performance which comes forward in bulk at the time it is required is of far greater value than a slightly superior machine two or three months too late. Once a design has been approved

in FRANCE by those responsible for using the machine, minor modifications and improvements in design must be rigorously forbidden, otherwise delays, such as have been experienced in the past as each new machine has been produced, will continue. Such delays result in our never knowing what we shall have available for any particular operations and in the consequent dislocation at the last moment of all our plans.

4. An important part of our aerial offensive next year will take the form of long distance bombing in GERMANY itself. This has been projected for the last two years as soon as resources permitted, and every precaution is being made to carry out the policy with the greatest vigour.

It is essential to remember, however, that long distance bombing can only be carried out on a sufficient scale if we are able, by pursuing our offensive policy elsewhere, to prevent undue interference with our bombing machines. It is of no use producing large numbers of bombing machines unless we can bring about the conditions which will admit of their full employment. The completion of the programme of fighting squadrons is therefore of primary importance and nothing should be allowed to interfere with the output of these machines.

5. It is presumed that the American Aviation will begin to appear on the front in the spring and summer of 1918. In view of the time which must elapse in aviation between the conception and the execution of a plan, it is very necessary to consider now the place the Americans can take in the general scheme of aerial operations on the Western front.

Apart from the necessary local work for their own Army which will require an organization of fighting, artillery, photographic, reconnaissance, and short distance bombing squadrons similar to our own and the French, American help as their production increases and in view of the enormous resources they should eventually possess, will be of especial value for bombing operations in GERMANY.

Assuming that the American Army will occupy its present situation in the line, the American aviation will be very favourably placed to bomb German industrial centres in the Upper RHINE Valley and should gradually allow us to withdraw our squadrons from the NANCY neighbourhood to our own Army area whence they should be able to bomb targets in the Lower RHINE Valley and beyond with the machines which should then be available. The advantages of operating from our own Army area are obvious.

If this policy is accepted the Americans should at once commence the preparation of the necessary aerodromes. The whole country consists of deep ridge and furrow, and though first-class aerodromes can be made on the tops of the hills, the work entailed is stupendous, and unless they begin at once there is no prospect of the aerodromes being ready before late next summer. Sites have, I understand, been allotted by the French, but no work whatever has been done on them. The aerodromes made by us will probably be available for the Americans in time, but our operations will have to continue for a long time after theirs commence.

By the time bombing by the Americans is in full swing it is more than likely that the enemy's opposition will have stiffened, so that the Americans will have to make provision as we are doing for fighting machines to accompany their bomb raids. As they will be operating, however, from their own Army area, and not from a far distant front as in our case, the fighting machines working on their Army front should be able to deal with any opposition in the forward areas, and it should only be necessary to provide a limited number of longer distance fighting squadrons to deal with any local protection the enemy may be induced to detail at the points of attack.

I would repeat that it is essential to get into touch with the Americans at once and decide upon our future joint policy in order to avoid a waste of resources on unfruitful production. Policy must in fact guide production instead of production dictating policy.

# APPENDIX XX

## FIGHTING IN THE AIR

*Memorandum issued by British G.H.Q., France, February 1918*

### I. GENERAL

1. *The Necessity of Fighting.*

The uses of aeroplanes in war in co-operation with other arms are many, but the efficient performance of their missions in every case depends on their ability to gain and maintain a position from which they can see the enemy's dispositions and movements. Cavalry on the ground have to fight and defeat the enemy's cavalry before they can gain information, and in the same way aerial fighting is usually necessary to enable aeroplanes to perform their other duties.

Artillery co-operation, photography, and similar work can only be successful if the enemy are prevented as far as possible from interfering with the machines engaged on these duties, and such work by hostile machines can only be prevented by interference on our part.

The moral effect of a successful cavalry action is very great; equally so is that of successful fighting in the air. This is due to the fact that in many cases the combat is actually seen from the ground, while the results of successful fighting, even when not visible, are apparent to all. The moral effect produced by an aeroplane is also out of all proportion to the material damage which it can inflict, which in itself is considerable, and the mere presence of a hostile machine above them inspires those on the ground with exaggerated forebodings of what it is capable of doing. On the other hand the moral effect on our own troops of aerial ascendancy is most marked, and the sight of numbers of our machines continually at work over the enemy has as good an effect as the presence of hostile machines above us has bad.

# FIGHTING IN THE AIR

## 2. *Similarity to Fighting on Land and Sea.*

To seek out and destroy the enemy's forces must therefore be the guiding principle of our tactics in the air, just as it is on land and at sea. The battle-ground must be of our own choosing and not of the enemy's, and victory in the fight, to be complete, must bring other important results in its train. These results can only be achieved by gaining and keeping the ascendancy in the air. The more complete the ascendancy, the more far-reaching will be the results.

The struggle for superiority takes the form, as in other fighting, of a series of combats, and it is by the moral and material effect of success in such combats that ascendancy over the enemy is gained.

## 3. *Necessity of Offensive Action.*

Offensive tactics are essential in aerial fighting for the following reasons:
(A) To gain the ascendancy alluded to above.
(B) Because the field of action of aeroplanes is over and in rear of the hostile forces, and we must, therefore, attack in order to enable our machines to accomplish their missions, and prevent those of the enemy from accomplishing theirs.
(C) Because the aeroplane is essentially a weapon of attack and not of defence. Fighting on land and sea takes place in two dimensions, but in the air we have to reckon with all three. Manœuvring room is, therefore, unlimited, and no number of aeroplanes acting on the defensive will necessarily prevent a determined pilot from reaching his objective.

## 4. *Choice of Objectives.*

An aerial offensive is conducted by means of:
(A) Offensive patrols.
(B) The attack with bombs and machine-gun fire of the enemy's troops, transport, billets, railway stations, rolling stock, and moving trains, ammunition dumps, &c., on the immediate front in connexion with operations on the ground.
(C) Similar attacks on centres of military importance at a distance from the battle front or in the enemy's country with a view to inflicting material damage and delay on his production and transport of war material and of lowering the moral of his industrial population.

### (A) *Offensive Patrols*

The sole mission of offensive patrols is to find and defeat the enemy's aeroplanes. Their normal sphere of action extends for some twenty miles behind the hostile battle line, and the farther back they can engage the enemy's fighting aeroplanes the more immunity will they secure for our machines doing artillery work, photography, and close reconnaissance. Since, however, aerial ascendancy will usually be relative only and seldom absolute, patrols are also required closer in to attack those of his fighting machines which elude the outer patrols, and to deal with his machines doing artillery observation and similar work. Fighting may take place at any height up to the limit to which the machine can ascend, known as its 'ceiling'. Artillery observation imposes a limit of some 10,000 feet, but fighting, bombing, and photographic machines may fly at any height up to 20,000 feet or even more. Offensive patrols must, therefore, work echelloned in height (see para. 10).

(B) *Attack of Ground Targets in the Battle Zone with Bombs and Machine-gun Fire*

The attack of ground targets cannot strictly speaking be described as fighting in the air, but it is an integral part of the aerial offensive designed to weaken the moral of, and cause material damage to, the enemy's troops. It is carried out by fast single-seater machines flying normally at anything from 100 to 2,000 feet, either singly or in formation. Fixed targets and, to a certain extent, troops can be attacked with advantage at any time including periods of sedentary warfare, but the attack of moving targets such as troops and transport is of the greatest value in connexion with ground operations either offensive or defensive.

(C) *Attack of Ground Targets at a Distance*

Targets at a distance are usually attacked by bombing. Such raids may be expected to produce their maximum effect when undertaken against distant objectives, since they may cause the enemy to withdraw artillery and aeroplanes from the front for the protection of the locality attacked. They are also, however, of great use in rear of the immediate front in connexion with operations on the ground.

Every patrol or raid should, therefore, be sent out with a definite mission, the successful performance of which will not only help us to gain aerial ascendancy by the destruction of hostile aircraft, but will also either tend to induce the enemy to act on the defensive in the air, or further the course of operations on the ground.

5. *Types of Fighting Machines.*

The machines at present in use for offensive purposes may be divided into four main classes:

(A) Fighters, (*a*) single-seaters, (*b*) two-seaters.
(B) Fighter reconnaissance.
(C) Bombers.
(D) Machines for attacking ground targets from a low altitude.

(A) *Fighters*

(*a*) *Single-seaters* are fast, easy to manœuvre, good climbers, and capable of diving steeply on an adversary from a height.

Their armament consists of two or more machine or Lewis guns, whose axis of fire is directed forward and, usually, in a fixed position in relation to the path of the machine.

Single-seater fighters are essentially adapted for offensive action and surprise. In defence they are dependent on their handiness, speed, and power of manœuvre. They have no advantage over a hostile single-seater

# FIGHTING IN THE AIR 95

as regards armament, and are at a disadvantage in this respect when opposed to a two-seater, and, therefore, the moment they cease to attack are in a position of inferiority, and must break off the combat, temporarily at any rate, until they have regained a favourable position. On the other hand, provided they are superior in speed and climb to their adversary, they can attack superior numbers with impunity, since they can break off the combat at will in case of necessity.

(b) *Two-seater fighters* have, in addition, a machine-gun for the observer on a mounting designed to give as wide an arc of fire as possible, especially to the flanks and rear. Their front gun or guns remain, however, their principal armament.

The two-seater is superior in armament to the single-seater, since it is capable of all-round fire, but is generally somewhat inferior in speed, climb, and power of manœuvre. It has greater powers of sustaining a prolonged combat, being less vulnerable to attacks from flanks and rear, but as in the case of single-seaters its chief strength lies in attack.

When fighting defensively or when surprised in an unfavourable position, it is often best for the pilot to fly his machine in such a way as to enable the observer to make the fullest use of his gun, while awaiting a good opportunity to regain the initiative.

### (B) *Fighter Reconnaissance Machines*

The first duty of these machines is to gain information. They do not go out with intent to fight, but must be capable of doing so, since fighting will often be necessary to enable the required information to be obtained. Those at present in use are two-seaters, the pilot flying the machine and the observer carrying out the reconnaissance. They approximate to the two-seater fighter type, and in the case of missions which can be carried out at 15,000 feet or upwards, are capable of acting alone, and usually do so.

### (C) *Bombing Machines*

Bombing machines usually carry at least one passenger, so that they can, in case of necessity, undertake their own protection, even when loaded. Their requirements, as regards armament, are similar to those of fighter reconnaissance machines. Machines carrying more than one passenger usually have a gun both fore and aft, and are strong for defensive fighting. The greater weight of bombs they can carry the better.

### (D) *Machines for Attacking Ground Targets*

Machines for this purpose will, as a rule, be single-seaters. Climb is of relatively minor importance, but they require to be fast and very manœuvrable and must have a very good view downwards. Single-seater fighters can be used for this work, but it is probable that a special type of machine will be evolved in which the pilots and some of the most vulnerable parts will be armoured. They will probably be adapted for carrying a few light bombs and will have at least one gun, capable of being fired downwards at an angle of 45° to the horizontal, and another firing straight ahead.

## FIGHTING IN THE AIR

### II. Principles of Aerial Fighting

6. *Factors of Success.*

The success of offensive tactics in the air depends on exactly the same factors as on land and sea. The principal of these are:
(A) Surprise.
(B) The power of manœuvre.
(C) Effective use of weapons.

#### (A) *Surprise*

Surprise has always been one of the most potent factors of success in war, and although it might at first sight appear that surprise is not possible in the air, in reality this is by no means the case. It must be remembered that the aeroplane is working in three dimensions, that the pilot's view must always be more or less obstructed by the wings and body of his machine, and that consequently it is often an easy matter for a single machine, or even two or three machines, to approach unseen, especially if between the hostile aeroplane and the sun. Fighting by single machines is, however, rapidly becoming the exception (see par. 7), and surprise is more difficult of attainment by machines flying in formation, though by no means impossible.

Even when in view, surprise is possible to a pilot who is thoroughly at home in the air, and can place his machine by a steep dive, a sharp turn, or the like in an unexpected position on the enemy's blind side or under his tail.

A surprise attack is much more demoralizing than any other form of attack and often results in the pilot attacked diving straight away or putting his machine into such a position that it forms an almost stationary target for a few seconds, and thus in either case affords the assailant an easy shot. To achieve surprise it is necessary to see the enemy before he sees you. To see other machines in the air sounds an easy matter, but in reality it is very difficult and necessitates careful training. The ground observer is guided by the noise of the engine, but the pilot, of course, hears no engine but his own. Again, while the ground observer sees the machine, broadly speaking, in plan, the pilot sees it in elevation, presenting a very much smaller surface. Add to these the variety of background, clear or cloudy sky, or the chequered appearance of the ground from above, and the obstruction offered to the pilot's view by the wings and fuselage of his machine, and the difficulties will begin to be realized.

Every pilot must, therefore, be trained to search the sky, when flying, in a methodical manner. A useful method is as follows: Divide the sky into three sectors by means of the top plane and centre section struts, and sweep each sector very carefully. From port wing tip to a centre section search straight ahead and then do the same from centre section to starboard wing tip. From starboard wing tip take a steady sweep straight upwards to port wing tip. In addition it is essential to keep a good lookout to the rear, both above and below the tail, in order to avoid being surprised. This can be done by swinging from side to side occasionally. The results of a concen-

# FIGHTING IN THE AIR

trated search of this description are surprising, while a pilot who just sweeps the sky at random will see little or nothing.

In addition to seeing the hostile machine it is necessary to recognize it as such. A close study of silhouettes will assist the pilots to do this, but until thoroughly experienced it is a safe rule to treat every machine as hostile. This, of course, necessitates going close enough to make sure, and soon results in a pilot becoming familiar with all types of machines in the air.

The types of hostile aeroplanes must be carefully studied, so that the performance and tactics of each, its blind side, and the best way to attack it, can be worked out. Some machines have a machine-gun mounted to fire downwards and backwards through the bottom of the fuselage.

Every advantage must be taken of the natural conditions, such as clouds, sun, and haze, in order to achieve a surprise.

If observed when attempting a surprise it is often best to turn away in the hope of disguising the fact that an attack is meditated. Flat turns may cause the enemy to lose sight of a machine even after he has once spotted it, as they expose much less surface to his view than do ordinary banking turns.

(B) *Power of Manœuvre*

Individual skill in manœuvre favours surprise, as pointed out above. Individual and collective power of manœuvre are essential if flying in formation is to be successful or even possible. They can only be obtained by constant practice.

To take full advantage of manœuvre the highest degree of skill in flying and controlling the machine is of the first importance. A pilot who has full confidence in his own powers can put his machine in any position suitable to the need of the moment, well knowing that he can regain control whenever he wishes. The best way to gain the required confidence is for the instructor to take his pupil up, dual control, throw the machine out of control himself, and allow the pupil to right it, the instructor only retaking control should the pupil fail to regain it. Once confidence has been acquired, practice will make perfect.

The second essential is that the pilot shall know his engine and how to get the best out of it, and thoroughly understand the use of his throttle. Many a chance is lost through pilots allowing their engine to choke in a dive, and no pilot can become really first-class unless he acquires complete practical familiarity with his engine by constant study and practice.

Good formation flying can only be carried out by pilots who know how to use their throttle. The leader must always fly throttled down or his formation will straggle, while they in turn must make constant use of their throttle to maintain station, and twist, turn, and wheel without confusion or loss of distance.

Other points to which attention must be paid are the following: Pilots must know the fuel capacity of their machine and its speed at all heights. The best height at which to fight varies with each type of aeroplane. Each pilot must know this height so that he can make the very best use of his machine. As a general rule machines should patrol at a greater altitude than their best fighting height. The direction and strength of the wind must be studied before leaving the ground and during flight. This study is most important since wind limits the range of action, as machines when fighting are bound to drift down wind.

### (C) *Effective use of Weapons*

(a) *Machine and Lewis Guns.* The essentials for successful fighting in the air are skill in handling the machine and a high degree of proficiency in the use of the gun and sights. Of these two essentials the second is of even more importance than the first. Many pilots who have not been exceptionally brilliant trick fliers have had the greatest success as fighting pilots owing to their skill in the use of the gun and sights. The manipulation of a gun in the air, especially on single-seater machines, is a very much more difficult matter than on the ground. Changing drums, for instance, though simple on the ground, is by no means easy when flying.

Every pilot and observer who is called upon to use a machine gun must have such an intimate knowledge of its mechanism as to know instinctively what is wrong when a stoppage occurs and, as far as the type of machine allows, must be able to rectify defects while flying. This demands constant study and practice both on the ground and in the air.

It is absolutely essential that pilots and observers should know exactly how their guns are shooting, and they should be tried on a target at least once a day. With his gun out of action a pilot or observer is helpless either for offence or defence.

Aerial gunnery is complicated by the fact that both gun and target are moving at variable speeds and on variable courses. Consequently, however skilful the flier, he cannot hope to be dead on the target for more than a very few seconds at a time, and it is essential that hand, eye, and brain be trained to work together.

Accurate shooting on the ground from a fixed gun at a fixed target is the first step in training; subsequently constant practice on the ground both when stationary and when moving at fixed and moving targets is essential. Finally, every opportunity must be taken of practice in the air under the conditions of a combat.

Except at point-blank range, it is essential to use the sights if accurate fire is to be obtained, and constant practice is needed with the sights provided. The aim can be checked with absolute accuracy by means of the gun camera, and combats in the air during which the camera is used are a most valuable form of training.

Tracer ammunition is of some assistance, but must be used in conjunction with the sights, and not in place of them. Not more than one bullet in three should be a tracer, otherwise the trace tends to become obscured. Too much reliance must not be placed on tracer ammunition at anything beyond short range. The principle should be to use the sights whenever possible at all ranges.

Inexperienced pilots are too apt to be content with diving and pointing their machine at the target and ignoring everything else. Mere noise and fright will not bring down an opponent; it is necessary to hit him in a vital spot. From the time a pilot starts to dive he should not have to fumble about for triggers and sights. His eye should fall automatically on the sight and his hand close on the trigger. By holding the right arm firmly against the body and working only from the elbow the machine can be held much steadier in a dive.

(b) *Bombs*. Skill and accuracy in bombing in the same way can only be acquired by continual practice and careful study of the conditions which govern the correct setting and use of bomb sights. Such practice is best obtained by the use of the Batchelor Mirror or of the camera obscura, and must be carried out from varying altitudes up to 15,000 feet, from which height bombs will often have to be dropped on service.

An exception must be made in the case of bombing by single-seater fighting machines from a low altitude, a method of attack which has been employed with very considerable success. In this case no sight is used, and the method found by experience to give the best results is to dive the machine steeply at a point on the ground a few yards in front of the target. The lag of a bomb released from a few hundred feet on a steep dive is very little. Individual pilots must find out by experiment exactly how far ahead they must aim.

### III. Formation Flying

*7. Evolution of Formation Flying.*

The development of aerial fighting has shown that certain fundamental maxims which govern fighting on land and sea are equally applicable in the air. Among these are concentration and mutual co-operation and support. The adoption of formation flying has followed as an inevitable result.

Any mission which has fighting for its object, or for the accomplishment of which fighting may normally be expected, must usually, therefore, be carried out by a number of machines, the number depending on the amount of opposition likely to be encountered and on a third fundamental axiom, namely, that no individual should have more than a limited number of units under his immediate control.

The evolution of formation flying has been gradual. When aerial fighting became general it was soon discovered that two machines working together had a better chance of bringing a combat to a decisive conclusion than had a single machine. The next step was for two or more pairs to work together, and this quickly became the accepted practice.

The chief difficulty is control by the leader of the remaining machines primarily due to the difficulty of communication in the air. For practical purposes this limits the number of machines that can be controlled by one man to six, and even when wireless telephony between machines is perfected this number is unlikely to be exceeded. The principles and causes which have led to formation flying remain in force, however, and are bound to result in a further development in the case of offensive fighting, namely two or more formations working in close co-operation with each other, and the best means of achieving such co-operation is the next problem to be solved in aerial warfare.

When a force on the ground is engaged in offensive action the troops comprising the main body must be protected from surprise from the front, flanks, and rear. Hence the universal employment in open warfare of advanced flank and rear guards. In the air the third dimension renders flank and rear guards unnecessary, their place being taken by the 'Above Guard' which can perform the duties of both. Whether we consider a single formation, therefore, or a group of formations acting in close co-operation, an 'Above Guard' is necessary and may consist of two or more machines in the first case or of one of the formations in the second. These should fly slightly above the main body, either directly behind or echelonned to a flank. The main body carries out the offensive fighting, the 'Above Guard' remaining intact above them to protect them from surprise.

8. *Some Principles of Formation Flying.*

The formations adopted vary in accordance with the mission and with the type of machine. Certain principles are, however, common to all formation flying and must be strictly observed.

(A) As on the ground so in the air the bed-rock of successful co-operation is drill, and good aerial drill is an essential preliminary to success in formation flying for any purpose. Before commencing drill in the air it has been found of great assistance to practise on the ground until all concerned are thoroughly conversant with the various evolutions. Simplicity is essential, and complicated manœuvres are bound to fail in action. Drill should commence in flight formation, each Flight Commander instructing and leading his own flight. Subsequently the Squadron Commander should lead and drill his whole squadron in three flights, each under its Flight Commander. A really well-drilled flight can manœuvre in the air with as little as a span and a half between wing tips, but in action it is better to keep a distance of 80 to 100 feet, otherwise pilots are apt to devote too much of their attention to avoiding each other.

(B) One of the first essentials of successful formation flying is that every pilot thoroughly understands the use of his throttle. He will have to use it constantly throughout the flight and must train himself to do so instinctively. The throttle must be used to keep station. If a pilot attempts to do so by sharp turns instead of by using his throttle he will inevitably throw the formation into disorder.

(C) The formation adopted must admit of quick and easy manœuvre by the formation as a whole.

(D) A leader must be appointed, and a sub-leader, in case the leader has to leave the formation for any reason, e.g. engine trouble. The machines of leaders and sub-leaders must be clearly marked. Streamers attached to different parts of the machine are suitable. Good formation flying depends very largely on the leader, who must realize that his responsibilities do not end with placing himself in front for others to follow. Their ability to do so depends very

# FIGHTING IN THE AIR 101

largely on himself and on constant practice together so that they know intuitively what he will do in any given circumstances.

(E) An air rendezvous must be appointed, and the leader must see pilots and observers before leaving the ground and explain his intentions to them. To save waste of time in picking up formation in the air and to ensure a really close formation, machines must leave the ground together or as nearly so as possible. When all machines have reached the rendezvous, the leader fires a signal light, indicating that formation is to be picked up at once. He should then fly straight for a short time, as slowly as possible, while his observer, if he has one, reports on the formation. If one or more machines are rather far behind, the leader should turn to the right or left, after he or his observer has given a signal that he is going to do so. Thus the machine behind will be enabled to cut a corner and close up. When the leader is satisfied with the formation he fires a light signifying that he is ready to start. The actual signal to start can be given either by the leader or from the ground; in the latter case the officer on the ground, who is responsible for the dispatch of the formation, will also be responsible for deciding when the proper formation has been adopted. It is usually best for the signal to be given from the ground. The decision as to the suitability or otherwise of the weather conditions will rest with the leader of the formation. A suitable code of signals for formation flying is given in Appendix A. Signal lights must be fired upwards by the leader, otherwise machines in the rear may have difficulty in seeing them.

(F) Pilots must clearly understand how the formation is to re-form after a fight. Once an attack has been launched it must tend to become a series of individual combats, but if a formation is able to rally at the first lull and make a second concerted attack it should gain a real advantage over a dispersed enemy formation. Definite instructions by the leader on the point are essential. A rendezvous over a pre-arranged spot has been found suitable, in the case of a small area. In the case of a large area two or more spots may be designated previously, the rendezvous to take place over the nearest. It must be realized that pre-arrangements may be found unsuitable, and in every case each pilot must invariably close on the nearest machine. If there is a choice he will join two machines in preference to a single machine and three machines in preference to two. This applies to the leader also. To rendezvous successfully after a fight needs continual practice.

(G) Formation must not open out under anti-aircraft gun fire. It has been found by experience that fire is usually less effective against a well-closed-up group of machines than when directed on a single machine. To open out is to give the enemy the chance, for which he is waiting, of attacking the machines of the formation singly. The enemy's aim can be thrown out temporarily, if the fire is very hot, by turning sharply, diving, or climbing, but it is seldom advisable to lose height, especially when far over the enemy's lines.

## FIGHTING IN THE AIR

9. *Use of Formation Flying.*

Flying in formation is necessary in the case of:
(A) Offensive patrols;
(B) Bomb raids;
and is the normal method of carrying out these duties.

Medium and long distance reconnaissances may also have to be carried out in formation, but a fast machine capable of flying at a great altitude can often carry out such reconnaissances by itself, including photography when large-scale photographs are not required. A further development of formation flying is in the attack of ground targets with machine-gun fire (see para. 12).

10. *Offensive Patrols.*

The sole duty of offensive patrols is to drive down and destroy hostile aeroplanes, and they should not be given other missions to perform, such as reconnaissances, which will restrict their fighting activities. In the face of opposition of any strength offensive patrols usually have to fly in formation in order to obtain the advantage of mutual support, but the formations adopted can be governed solely by the requirements of offensive fighting. Single-seater scouts or even two-seaters, if superior in speed and climb to the great majority of the enemy's machines, may be able to patrol very successfully alone or in pairs, taking advantage of their power of manœuvre and acting largely by surprise, but in the case of machines which do not enjoy any marked superiority, formation flying is essential. Fighting in the air, however, even when many machines are involved on each side, tends to resolve itself into a number of more or less independent combats, and it has been found advisable to organize a purely fighting formation accordingly. Such a formation can suitably consist of six machines, organized in groups of two or three machines each, every group having its own sub-leader, the senior of whom takes command of the formation. A deputy leader should also be designated, in case the leader falls out for any reason. As far as possible the groups should be permanent organizations, in order that the pilots may acquire that mutual confidence and knowledge of each other's tactics and methods which is essential for successful fighting. It must be impressed on pilots that the group is the fighting unit and not the individual (see para. 14).

11. *Reconnaissances and Bomb Raids.*

In reconnaissance the whole object is to protect the reconnaissance machine or machines, and enable them to complete their work. Opposition will usually take one of two forms. The enemy's scouts may employ guerrilla tactics, hanging on the flanks and rear of the formation, ready to cut off stragglers, or attacking from several directions simultaneously; or else the formation may be attacked by a hostile formation. The modern type of two-seater fighter reconnaissance machine is able to deal with either class of opposition without assistance. The machines must fly in close formation, keep off enemy scouts which employ guerrilla tactics by

# FIGHTING IN THE AIR

long range fire, and be ready to attack a hostile formation if the enemy's opposition takes that form.

Reconnaissance formations, like fighting formations, can be organized in groups, each with its sub-leader, but as the object is to secure the safety of the reconnaissance machine, the whole formation must keep together and act as one.

A suitable formation in the case of six two-seater machines has been found to be two lines of three, the flankers in the front being slightly higher than the centre (reconnaissance) machine, and the three machines in rear slightly higher again. The intervals between the machines should not be more than 100 yards, and the distance of the rear rank from the front should be sufficient only to admit of a good view being obtained of the leading machines.

The pace must be slow, otherwise the rear machines are bound to straggle. Machines must, therefore, fly throttled down. Sharp turns by the leader also lead to straggling; a signal, therefore, should always be given before turning, and a minute or two allowed, if possible, after giving the signal before the turn is commenced, in order to give the machines on the outer flank time to gain ground.

The duty of the bombing machines is to get to their objective and to drop their bombs on it and only to fight in the execution of their duty. The secret of success is the most careful pre-arrangements, so that every one knows exactly what he has to do. The bombing machines, like a reconnaissance, must keep in close formation. Any tendency to straggle or to open out under anti-aircraft fire will give the enemy the opportunity he is seeking to attack and split up the formation. A well-kept formation, on the other hand, is seldom attacked at close range, unless by very superior numbers. When bombing from a height the best results have been secured by dropping bombs while still in formation. Three machines drop their bombs simultaneously, the centre observer being responsible for the sighting or, if preferred, all machines can drop their bombs simultaneously on a signal from the leader. If it is necessary for machines to break formation to drop their bombs, a rallying-point must always be chosen beforehand where they will collect and resume flying formation as soon as their bombs have been released.

When a very large raid is contemplated, it will often be best to carry out the attack by two separate formations, since there is a limit to the number of machines which can be controlled efficiently by a single leader. Six bombing machines are normally the maximum. The departure of the two formations from their respective rendezvous, if they are to make a single raid, should be so arranged as to enable them to give one another mutual support in case of a heavy hostile attack. The rendezvous should not be too close together, 10 to 15 miles apart is a suitable distance. Departures from the rendezvous should be timed so that the first formation is leaving the objective as the second approaches, and the leaders should watch each other's signals.

With modern machines an escort to a reconnaissance formation or bomb raid is seldom desirable, and far better results are obtained by sending one or more offensive patrols to work independently over the area where opposition to the reconnaissance or raid is most likely to be encountered. If an escort is provided, its primary duty is to enable the reconnaissance or raid to accomplish its mission and it should only fight in the execution of this duty. It is usually best to keep the escort and the machines it is protecting as distinct formations under a separate leader. The escort flies above the reconnaissance or bombing machines in such a position as to obtain the best view of them and the greatest freedom of manœuvre in any direction. Its role is:

(A) To break up an opposing formation.
(B) To prevent the concentration of superior force on any part of the formation they are protecting.
(C) To assist any machine which drops out of the formation through engine or other trouble.

While the bombs are being dropped, the escort should circle round above the bombing machines, protecting them from attack above, and ready to dive on to any hostile machine that may interfere with them.

12. *Attack of Ground Targets.*

Formation flying has lately been adopted for the attack of ground targets with excellent results, formations appearing to be no more vulnerable to rifle and machine-gun fire from the ground than is a single machine. This is probably due to a tendency to fire at the formation as a whole instead of picking out a particular machine. On the other hand, a formation, as against a single machine, possesses the following advantages:

(A) There is less chance of machines losing their way as there are several individuals instead of one only attempting to keep their bearings.
(B) A greater volume of fire is brought to bear on any target discovered.
(C) A formation is stronger if attacked.
(D) A formation may be expected to have greater moral effect on the enemy's troops.

Formation flying at low altitudes demands even more constant practice together than does formation flying at a height, because fire from the ground makes continuous changes of direction and height a necessity. A suitable height from which to attack ground targets is 600 to 800 feet. The essential point is to go low enough to make certain of differentiating between our own and the enemy's troops. Above 800 feet this is difficult, and the chance of interference by hostile aircraft is greater, but these seldom come down to fight below 1,000 feet. Formations for low flying should never exceed six machines.

## IV. Fighting Tactics

13. *General.*

Fighting tactics vary with the type of machines and with the powers and favourite methods of individual pilots. No hard-and-fast rules can be laid down, but the following hints based on the experience of others may be of use to the young pilot until he has acquired experience of his

# FIGHTING IN THE AIR

own. There are four golden rules which are applicable to all offensive aerial fighting:
- (A) Every attack must be made with determination and with but one object, the destruction of the opponent.
- (B) Surprise must be employed whenever possible.
- (C) If surprised or forced into an unfavourable position a pilot must never, under any circumstances, dive straight away from his opponent. To do so is to court disaster, since a diving machine is an almost stationary target. Moreover, the tactical advantage of height is lost by diving and the initiative surrendered to the hostile machine. The best course of action depends on the type of machine and is discussed below.
- (D) Height invariably confers the tactical advantage.

## 14. *Single-seater Fighting.*

Fighting in formation with single-seaters is a most difficult operation and demands constant study and practice, the highest degree of skill on the part of the individual pilots, mutual confidence between them, and intimate knowledge of each other's methods.

The patrol leader's work consists more in paying attention to the main points affecting the fight than in doing a large share of the fighting himself. These main points are:
- (A) The arrival of more hostile machines, which have tactical advantage, i.e. height.
- (B) The danger of the patrol being carried by the wind beyond the range of its petrol supply.
- (C) The patrol getting below the bulk of the hostile formation.

As soon as any of these conditions occur it is usually better to break off the fight temporarily, and to rally and climb above the enemy before attacking them again.

When fighting in formations of two or more groups, the fighting unit should be the group, each selecting its own objective and acting as described below. The groups will often become separated, but every effort should be made to retain cohesion within the groups. The practice of individual pilots breaking away from the formation to attack hostile machines almost always leads to disaster sooner or later. If the enemy machines scatter, attention should be concentrated on those lagging behind, and, if they dive and are followed down, at least one group should remain at a height as a protection from surprise. The dangerous quarter in the case of a formation of single-seaters is the rear, and care must always be taken to keep a constant watch behind and above.

If surprised in an unfavourable position it should be the invariable rule, if time permits, to turn and attack the adversary before he comes to close quarters. If, however, he succeeds in doing so, the best chance lies in a quick climbing turn. Any method which entails losing height such as a side-slip or a spin is bad, as the hostile machine has merely to follow and attack afresh from above.

Surprise by a formation is difficult, and success must be sought in close co-operation and boldness of attack. If the enemy is inferior in numbers an opportunity will occur for a concerted attack by a group against a single machine. The actual attack should be carried out by two machines, the third remaining above to protect them from surprise. The two attacking machines may converge on the enemy from different directions on the same level, but the attacks must be simultaneous so that they cannot be engaged separately. Another method is to attack echelonned in height, the lower machine diving and attacking the enemy from behind, while the upper machine awaits an opportunity to swoop down on him when he turns to engage the machine that attacked first.

An attack of equal numbers will usually resolve itself into a series of individual duels. The leader must always ensure that his formation is well closed up before attacking, giving the rear machine time if necessary so that all pilots can attack their adversaries simultaneously.

In attacking superior numbers the best chance of success lies in the destruction of the enemy's morale by excessive boldness.

Decoy tactics are sometimes successful. One group attempts to draw the enemy on to attack, while the other flies high above it, ready to surprise the enemy should he seize the apparent opportunity. Watch must be kept for similar tactics on the part of the enemy.

The group going down as a decoy must not be more than about 3,000 ft. below the remainder or it will run the risk of being attacked from the flank by superior numbers before the groups above can get down to its assistance.

If, owing to being cut off from his formation and being attacked by a superior number of machines, a pilot is forced down low, his best method of escape is usually to go down quite close to the ground and fly back on a zigzag course.

Although as a principle single-seaters should not act alone, yet in many cases isolated scouts will be called upon to fight single-handed, e.g. when a formation has become split up during a combat and a machine fails to rejoin its formation. Again, selected pilots on the fastest types of single-seaters may be usefully employed on a roving commission, which will enable them to make the greatest use of surprise tactics.

Single-seater fighting calls for much individual initiative especially when a combat develops itself into individual fighting and the pilot has the opportunity of developing his own particular method of attack. Methods vary with the type of machine attacked, and may be conveniently discussed under two headings:

(*a*) Single-seater against single-seater.

(*b*) Single-seater against machines with one or more passengers.

(*a*) *Single-seaters are best attacked from above* and behind with a view to getting within point-blank range if not observed. Height enables the attacker to anticipate his enemy's movements more quickly and to guard himself from attack from behind by a sudden turn on the part of his opponent. It is therefore essential to have plenty of engine power in hand so as to keep the means of climbing above the enemy throughout the fight and thus retain the advantage of height whatever tactics he may

# FIGHTING IN THE AIR 107

pursue. When attacking a hostile formation, one of their number, more often than not their leader, will sometimes fly out of the fight and climb his utmost with a view to getting above the attackers. The leader of the attacking formation should watch for this manœuvre, and be ready to frustrate it by climbing himself. The knowledge that there is one enemy above not only nullifies the advantage in height but divides the attention of the attacking pilots just when it should be entirely concentrated on the machines they have severally selected to attack.

A hostile pilot who attempts to come up unawares from behind and below can usually be defeated by a quick climbing turn. He will often be taken by surprise and turn flat, offering a vulnerable target to attack from above.

Attacks from directly in front or from the flanks are often successful, as the vital parts of the machine from the pilot forward are fully exposed. Aim should be taken at the front of the machine in such an attack. It is a common mistake to aim at the pilot, which usually only results in hitting the fuselage, as the majority of the fire usually takes effect behind the point of aim. This is conclusively proved by the number of our machines which return with the fuselage riddled and little or no damage from the pilot forward.

Similarly, when attacking from above and behind, aim should be taken at the leading edge of the top plane, thus increasing the chance of hitting the engine and pilot.

When it is necessary to swerve to avoid a collision or to break off the combat temporarily to change a drum or rectify a jamb, this should be done by a sudden turn or climb, care being taken subsequently to avoid flying straight or losing height. When ready, a favourable position must be regained by manœuvre before renewing the attack.

(b) *Single-seaters attacking two-seaters* can do so from behind and above, from behind and below, or from front or flanks. The most favourable method is perhaps to attack from behind and below, attempting to achieve surprise by climbing up under the fuselage and tail plane, the blindest spot from the point of view of the observer. A skilfully handled single-seater which can attain a position about 100 yards behind and 50 feet below a hostile two-seater without being observed, is in a position to do most damage to the enemy with least risk to himself. Once in this position the object of the attacker must be to keep out of the enemy's field of fire as much as possible. The two-seater will endeavour to bring fire to bear on the attacker by turning quickly in order to deprive him of the cover of the fuselage, and great skill is required to retain a position directly in rear in spite of frequent turns. If enjoying superior speed, which will usually be the case, the single-seater should turn always in the opposite direction to the two-seater, e.g. if the two-seater turns to the right, the attacker at once turns to the left, thus preserving their relative positions. When on the bank in the act of turning, the two-seater will offer a favourable target to the attacker if the latter is quick enough to take advantage. A short quick burst at this moment may confuse the pilot and cause him to dive, in which position it will be very difficult for the observer to do any accurate shooting, or even to stand up to fire, owing to the wind pressure, and it is safe to disregard the rear gun for the time being. Should the observer be put out of action the rear gun can, of course, be disregarded altogether and the attacker can close to point-blank range.

When attacking two-seaters from above a short-steep dive is effective, because the gunner has then to shoot almost vertically upwards, which is difficult and impairs the accuracy of his aim. If approaching head on with a view to turning and attacking from behind, the turn must be made before a position vertically over the opponent is reached, otherwise the attacker will be left two or three hundred yards behind the hostile machine with no chance of surprise and in not a very favourable position for attack. An attack from the front and above or from the flanks precludes the use of the observer's gun altogether in many types of machines, but care must be taken not to give the observer an easy shot by diving straight on past the machine after delivering the attack.

Surprise can often be attained by carefully watching the adversary, preferably from behind. An especially favourable opportunity for surprise occurs in the case of a hostile machine crossing our front on some special mission, for once the hostile observer has satisfied himself that the air is clear, he will give his principal attention to his work. The enemy will often choose cloudy weather for such missions, and this gives special chances of surprise to a skilful pilot, working with intelligence. In such weather it must be remembered that it is often of advantage to approach the hostile machine on his own level when the planes form but a thin line which is difficult to see. When surprise is impossible, advantage must be taken of the handiness and manœuvring power of the scout to prevent the enemy from taking careful aim by approaching him in a zigzag course, and never in a straight line, since a machine attacking in a straight line offers a comparatively easy target. When within about 100 yards the zigzag course must be abandoned, and the moment when the enemy is in the act of shifting his aim should, if possible, be chosen. He can then be attacked in a straight line with a burst of rapid fire, or it may be possible to get below him and fire at him more or less vertically at almost point-blank range.

To open fire at long range is to give the advantage to the enemy, since it is necessary to fly straight to bring fire to bear, and an easy mark is thus offered.

In the case of a group attacking a single two-seater, as in that of single-seaters, one machine must remain as an 'Above Guard'. The other two will have a very good chance of surprise if one machine repeatedly makes short dives, firing a few rounds and climbing again. This will engage the attention of the observer and afford the second machine an opportunity of creeping up underneath the enemy to point-blank range.

In the attack of multi-seater machines, surprise is even more essential to success, since they usually have a gun on a circular mounting both in front and rear, and consequently have practically no blind spot. Some types have also a gun mounted to fire downwards at an angle through the fuselage in order to deal with attack from behind and below.

## 15. *Two-seater Fighting.*

The principles of fighting in two-seaters designed for the purpose are similar to the above, but in the actual combat they are able to rely more on their power of all-round fire and less on quickness of manœuvre. The fighting tactics adopted should, therefore, be such as to favour the development of fire. The single-seater, when no longer able to approach its adversary, temporarily loses all power of offence and has to manœuvre to regain a favourable position. The two-seater, on the other hand, can develop fire from its rear gun, after passing its adversary or on the turn. The gun or guns firing straight ahead must be looked on as the principal weapons, the fire of the observer being brought to bear after passing the adversary, on a turn, or against another machine attacking him from the rear.

A two-seater, like a single-seater, must, however, never dive straight away from an adversary, as even though it can fire to the rear the advantage is all with the machine which is following.

Formations of two-seaters are less liable to surprise from the rear, since the observers of the rear machines can face in that direction and keep a constant look-out. Mutual fire support is also easier in their case, in view of their all-round fire. They are, therefore, as already pointed out, better able to sustain a protracted battle. The essence of successful fighting in two-seaters lies in the closest co-operation between pilot and observer. They must study their fighting tactics together, and each must know what the other will do in every possible situation.

The tactics of an artillery or bombing machine should be more defensive in their nature since their primary work is not to fight but to fulfil their mission. Machines of these types are also usually at a considerable disadvantage as regards quickness of manœuvre. They should therefore be fought in such a way as to give the observer every chance of bringing effective fire to bear, and the front gun should be retained for use when opportunity offers, such as when a hostile machine attacking from behind overshoots the mark.

## 16. *Fire Tactics.*

Opportunities in the air are almost invariably fleeting, and consequently the most must be made of them when they occur. Fire should, therefore, be reserved until a really favourable target is presented, and should then be in rapid bursts. Fire should only be opened at ranges over 300 yards when the object is to prevent hostile machines coming to close quarters, as in the case of an escort to a reconnaissance machine, and should not be opened at ranges over 500 yards under any circumstances. In offensive fighting the longer fire can be reserved, and the shorter the range, the greater the probability of decisive result.

For an observer on a two-seater machine, however, a range of from 200 to 300 yards is suitable, since it enables full advantage to be taken of the sight. Fire may be opened at longer range when meeting a hostile machine than when overhauling it, otherwise there will be no time to get in more

# FIGHTING IN THE AIR

than a very few rounds owing to the speed with which the machines are approaching one another. Pilots and observers must accustom themselves to judging the range by the apparent size of the hostile aeroplane and the clearness with which its detail can be seen. This needs constant practice.

A reserve of ammunition should be kept for the return journey when fighting far over the lines.

Manœuvre is an integral part of fire tactics, and every endeavour must be made to manœuvre in such a way as to create favourable opportunities for one's own fire and deny such opportunities to the enemy.

## APPENDIX A

*Code of Light Signals to be used in Formation Flying.*

| Colour. | Fired by. | Indicates. |
| --- | --- | --- |
| Red | Leader, in conjunction with K strips and red light from ground. | Leave the rendezvous. Leader fires a light to indicate he is ready to leave the rendezvous. The formation leaves on this signal or awaits an order from the ground consisting of K and a red light. |
| White | Leader, or from the ground in conjunction with N in strips. | Return to your aerodrome. Expedition abandoned. This signal applies east or west of the line. If fired east of the line it also indicates 'keep formation till line is crossed'. |
| Red | Any member of formation. | 'I am being attacked and need assistance.' |
| Red | Leader. | Rally to continue operation—(attack having been dispersed). |
| Green | Any member of expedition (including leader). | 'I am forced to return to my aerodrome.' This signal, if fired by the leader, does not imply that the expedition is abandoned. The leadership must be taken up by the deputy leader. |

## APPENDIX XXI

### BOMBING OPERATIONS

*Memorandum submitted to G.H.Q. by Major-General J. M. Salmond, G.O.C. Royal Air Force, France: June 1918*

G.H.Q.

1. I understand it is proposed to adopt as far as possible a policy of concentrating our bombing operations against selected points on the enemy's communications with a view to material effect. Such a policy depends for success on:

(*a*) the means at our disposal;
(*b*) a suitable organization;

# BOMBING OPERATIONS

(c) the selection of really important objectives;
(d) to what extent material damage can be caused by bombing: (i) by day and (ii) by night.

2. As regards the means at our disposal for day bombing, Corps machines and Scouts can be disregarded as they must work at comparatively short distances over the lines, since they are not capable of attacking targets at any distance. There remain, therefore, the fighter reconnaissance squadrons, of which we have at present ten, which, when the 9th Brigade returns on the 21st instant, will be distributed as under:

|              |   |
|--------------|---|
| First Army   | 2 |
| Second Army  | 2 |
| Third Army   | 2 |
| Fourth Army  | 1 |
| G.H.Q. Brigade | 3 |

G.H.Q. and Armies must have at least one such squadron each for long-distance reconnaissance and photography, so that the number available for day bombing is at present 5.

3. As regards night bombing, we have at present 7 squadrons distributed as under:

|               |        |
|---------------|--------|
| Armies        | 1 each |
| G.H.Q. Brigade | 3     |

An additional squadron is due at the end of the month and is to join the G.H.Q. Brigade.

4. As regards organization, centralized control necessitates the assembly of all the squadrons concerned in a relatively small area. Squadrons which are dispersed in the various Army areas cannot be controlled from H.Q., R.A.F. Before sending out a raid information has to be obtained from the Army concerned as to the weather on its front, and on a doubtful day the opportunity has probably passed before this has been done and the order to start has been given. Central control of dispersed squadrons by means of a time-table has been suggested, but this is not practicable in aerial work which is so dependent on the weather, and would inevitably result in less work being accomplished, since a squadron might very well not be able to get out at the time allotted to it, and would not under a system of central control be able to go out at another time in the day when weather was suitable, while it could do so if controlled locally.

On the other hand, if our efforts are to be concentrated on a few points each day, central control is very desirable for day bombing and essential for night bombing.

5. Our effective bombing force is limited. The selection of objectives presents a good deal of difficulty. Before a battle we must hamper the enemy's preparation and concentration, but when we are on the defensive it is always a matter of difficulty to locate the point of attack. Except during a concentration, it is doubtful to what extent any damage we can hope to inflict would cause any real dislocation of the enemy communications. During a battle all available resources must be used to enable us to keep our artillery machines in the air. If this object is attained, any surplus

machines should be used to bomb detraining points, debussing centres, close railway junctions.

With the limited bombing units at our disposal there will be no *effective* bombing force available *during a battle* to undertake long expeditions to distant objectives.

Again, during the day it is quite possible to employ a large number of squadrons against one objective, but this cannot be done during the night owing to the real or supposed danger of collisions. Even if this danger is not very real, the fear of collision is bound to react on pilots, and for this reason I doubt if more than one squadron could be allotted to each target, and concentration would therefore be difficult and night bombing is unfortunately the best opportunity for material effect.

6. Material damage from day bombing is, I am afraid, very small and must remain so as long as it is necessary to bomb from great heights at which an error of 1,000 yards is not at all excessive. Material damage from night bombing is undoubtedly greater on suitable nights, but all experience in this war shows that it is very seldom vital.

There were over 100 night and day raids on Dunkirk last summer and autumn, but we have continued to use the port without interruption.

Before the SOMME Battle in 1916 we selected six points, the temporary destruction of which—if it could have been secured—would have isolated the battle area from all reinforcements. At only one did we achieve any marked success, i.e. at ST. QUENTIN, where infantry were caught entraining and an ammunition train blown up, and the total effect as regards holding up reinforcements was practically nil as far as we know.

7. It seems to me that if all our resources for day and night bombing are centralized, we shall lose a good many opportunities of inflicting moral and material damage on the enemy in rest billets, during reliefs, at forward railheads, dumps, &c., for a very doubtful advantage, and I would therefore suggest that each Army should retain one day and one night bombing squadron to deal with such targets as the above, and that all squadrons over and above these should be concentrated under H.Q., R.A.F., to take temporary control of the bombing squadrons of the Army on whose front the G.H.Q. squadrons were operating. This would give us for day bombing:

(*a*) 2 G.H.Q. squadrons,
(*b*) 1 squadron from the Army concerned,
(*c*) any additional squadrons that may be temporarily deflected to the
   E.F. from the Independent Force, probably, I understand, 2 or 3;

and for night bombing:

(*a*) 3 squadrons of which one is a Handley Page,
(*b*) a second Handley Page squadron due in a week's time,
(*c*) 1 squadron from the Army concerned.

These squadrons to be used for the purpose of, if possible, isolating a particular front.

The squadrons remaining with armies not immediately concerned should, as circumstances arose, be directed from G.H.Q. to direct their efforts on targets on their own front which might affect the particular case.

With the above force a fairly concentrated bombing programme could

# BOMBING OPERATIONS

be carried out against a strictly limited number of targets, not exceeding four, and we should then be able to judge by results how far the system of concentration to achieve material damage is practically sound.

H.Q., R.A.F.
20th June 1918.

## APPENDIX XXII
### PROTECTION AGAINST ENEMY AEROPLANES
*Translation from a German document, July 1918*

1. During our offensives, losses through the action of enemy aviators have proved to be extraordinarily high. In order to avoid such losses in future, and to accustom the troops to take advantage of effective methods of protection against aviators, measures must be taken during quiet periods continuously to improve the protection of the troops against enemy air attacks.

During operations on a large scale, the available anti-aircraft units, even though working in close co-operation with aircraft units, will not be able to undertake continuous protection of the troops, more especially in rear areas, although they will be of undeniable value. Anti-aircraft defence on quiet fronts naturally shows still wider gaps. Above all, protection during quiet periods is necessary in rear areas against enemy long-distance raids. The troops must themselves create this with their own machine-guns, and organize their own arrangements for interference with the enemy's intentions.

2. To ensure the safety of the troops on the march (especially at night) and during training, and the prevention of enemy air reconnaissance, the following points must be noted:

   (i) Avoid main roads; march as far as possible on by-roads and in small columns. In order to avoid crowding and blocks, every march must be arranged according to a regular time-table, and allotted a definite route. Troops must not collect at stations, or in villages.

   (ii) A look-out for enemy machines must be kept at every halt and during the march itself. On the approach of enemy aviators, troops must clear the road and get under cover from air observation (in ditches, or in groups underneath the trees). Horses and vehicles must, if possible, be drawn up close underneath the trees on one side of the road. No movement. If the route is illuminated at night by parachute flares, halt immediately, clear the road, lie down. Pull in vehicles and horses close to the trees.

   (iii) All troops provided with machine-guns, and especially the machine-guns detailed for anti-aircraft defence during the march, must come into action; the latter should be mounted upon the vehicles, or otherwise disposed, so as to be immediately ready for action.

   (iv) At all training exercises, at reviews, and in assembly positions, especially during fine weather, lookouts and machine-guns are to be installed for observation and protective purposes. Horses must not be grouped in large numbers. Vehicles, when halted, must be placed under cover or dispersed.

Artillery must take with them to all practices the machine-guns allotted for anti-aircraft defence.

(v) Ammunition columns are already provided with machine-guns for their protection. Baggage and other transport columns, when moving by march route, will also be allotted machine-guns.

The order goes on to prescribe further detailed measures of protection, among them the following:

*In billets*, entrances of all suitable cellars will be marked by white luminous figures showing the accommodation available.

Zigzag trenches will be dug round all houses, huts, and tents, and their location shown by luminous arrows on the sides of the buildings.

Every locality will be provided with sufficient dug-out accommodation for the whole of the troops quartered in it.

*Along the main roads*, short lengths of zigzag trenches, running obliquely from the side of the road, will be made at intervals of 100–200 metres, especially in areas where an offensive is intended.

*During a battle*, machine-gun nests echelonned in depth must be organized in each divisional sector.

*In conclusion*, the order calls attention to the necessity for strict fire discipline and a good lookout system.

Aeroplanes flying at over 3,000 feet are not to be fired at.

## APPENDIX XXIII

### METHODS OF BOMBING

*Report from the Experimental Station, Orfordness, October 1918*

General.

A number of reports on methods of bombing have been written recently by Orfordness as the result of experimental work; this report has been written in order to show various methods in stronger contrast than they appear when published separately.

Each method is stated quite shortly and where possible the Orfordness Report dealing with the subject is given.[1] It is hoped that this report will prevent any confusion of ideas which might occur due to Reports being published at different times.

A. *Day Bombing.*

1. To obtain the best results from a number of machines at great heights, bombing should be done in formation; and this is quite independent of such problems as defence, &c. Evidence of this can be obtained from Orfordness Reports: when dropping in formation the 50 per cent. total

---

[1] *Author's note.*—References omitted.

## METHODS OF BOMBING 115

error was $3\frac{1}{2}°$ and when each machine was dropping individually, the 50 per cent. total error was $4\frac{1}{4}°$ (1° from 15,000 feet is equal to 88 yards).

For high bombing, the formation should drop on a signal from the leader's machine. The leader's observer should use a bomb sight of the High Altitude Drift or Double Ball Level type. The leader's observer should be responsible for steering the leader over the target and should pull off the bombs. Duplicate releases should be fitted to the other machines, so that either pilot or observer can pull off the bombs on the leader's signal.

2. For low bombing with two-seater machines each machine should be provided with a C.F.S. 7 sight for the pilot. The best results are obtained if the sighting and releasing of the bombs is done by the pilot without any assistance from the observer.

3. For low bombing, from single-seater machines, the best method is for the pilot to dive at the target; no sight is required, but a certain amount of practice by each pilot is necessary so that he may get used to the angle which he must allow for different speeds of dive.

B. *Night Bombing.*

1. For heavy bombing machines of the Handley Page type, the most satisfactory sight is the H.A.D. or the Course Setting Bomb Sight, when supplies of the latter are available. The best method is for the observer to steer the pilot and release the bombs. Considerably greater accuracy can be expected at night with a sight which does not limit the direction of attack to an up or down wind direction.

2. From smaller night bombing machines the bombs may either be dropped by the observer, while the machine is flying level, using a C.F.S. 7 sight, or the pilot may dive at the target and the observer release the bombs on the dive. The latter method is popular with F.E.2b machines, as the pilot is able to see the target over the front of the machine, when it is diving, and very fair accuracy may be obtained if the pilot and the observer have had some practice together by day.

*Railway Bombing.*

In the above notes it has been assumed that a target such as a town, factory, station, &c., is being attacked; it may, however, be important to cut a railway line. A large number of tests have been made attacking the line by either flying along it or across it; these, coupled with the Medbourne Trials, all go to show that the bomb has to fall so close to the line to do any damage, that it is impracticable to attempt to cut railway lines by high bombing. The only satisfactory method appears to be to fly along the track low down at night with a pusher type machine such as an F.E.2b.

*Cloud Flying.*

It is important that bombing pilots be trained to fly on a definite compass course in clouds. This will enable raids to be made on days when there are scattered clouds and the gaps are not large enough for a formation to be able to climb through.

## METHODS OF BOMBING

On days when there is a continuous layer of clouds low down, it is possible to fly in the clouds until over the objective, when the pilot can dive down, drop the bombs, and climb back into the clouds before the enemy can retaliate.

Orfordness.
17.10.18.

# APPENDIX XXIV

## ORDER OF BATTLE OF THE ROYAL AIR FORCE, FRANCE, ON 8th AUGUST 1918

*General Officer Commanding.* Major-General J. M. Salmond, C.M.G., C.V.O., D.S.O., St. André-aux-Bois

*H.Q. Communication Squadron.* Capt. J. C. Liddle (A.W., R.E.8, B.E.2d and e) . Berck-sur-Mer

*Total aeroplanes on charge* . . 15

*H.Q. IX (H.Q.) Brigade.* Brig.-Gen. R. E. T. Hogg, C.I.E. . . . . . St. André-aux-Bois
*Squadron* 25 (D.H.4) Maj. C. S. Duffus, M.C.  Ruisseauville

*Total aeroplanes on charge* . . 18

*H.Q. Ninth Wing.* Lieut.-Col. A. V. Holt, D.S.O. . . . . . . Mondicourt.
*Squadrons.*
   27 (D.H.9) Maj. G. D. Hill . . Beauvois
   32 (S.E.5a) Maj. J. C. Russell . . Bellevue
   49 (D.H.9) Maj. B. S. Benning . . Beauvois
   62 (Bristol Fighter) Maj. F. W. Smith . Croisette
   73 (Sopwith 'Camel') Maj. M. le Blanc-Smith, D.F.C. . . . Bellevue

*Total aeroplanes on charge* . . 99

*H.Q. Fifty-First Wing.* Lieut.-Col. R. P. Mills, M.C. . . . . . Croisette
*Squadrons.*
   1 (S.E.5a) Maj. W. E. Young, D.F.C. . Fienvillers
   43 (Sopwith 'Camel') Maj. C. C. Miles, M.C. . . . . . . Fienvillers
   54 (Sopwith 'Camel') Maj. R. S. Maxwell, M.C. . . . . . . Fienvillers
   98 (D.H.9) Maj. E. T. Newton-Clare, D.S.O. . . . . . Blangermont
   107 (D.H.9) Maj. J. R. Howett . . Ecoivres

*Total aeroplanes on charge* . . 101

## ORDER OF BATTLE—AUGUST 1918

H.Q. *Fifty-Fourth Wing.* Lieut.-Col. R. G. D.
Small . . . . . . . Lambus
*Squadrons.*
   58 (F.E.2b) Maj. D. Gilley, D.F.C. . Fauquembergues
   83 (F.E.2b) Maj. S. W. Price . . Franqueville
   151 (Sopwith 'Camel') Maj. C. J. Q. Brand,
    D.S.O., M.C. . . . . . Fontane-sur-Maye
   207 (Handley Page) Maj. G. R. Elliott . Ligescourt
   215 (Handley Page) Maj. J. F. Jones,
    D.S.C. . . . . . . Alquines
   'I' Flight (F.E.2b) Capt. P. W. B. Law-
    rence . . . . . . Fauquembergues
   *Total aeroplanes on charge* . . 76

Ninth Aircraft Park. Maj. A. F. Palmer . Vacquerie-le-Boucq
H.Q. *I. Brigade.* Brig.-Gen. D. le G. Pitcher,
  C.M.G. . . . . . . Chateau Tenby
H.Q. *First (Corps) Wing.* Lieut.-Col. E. L.
  Gossage, M.C. . . . . . Hill 180 (1 mile N.E.
                                     of Baraffle)

*Corps.*   *Squadrons.*
I     2 (A.W.) Maj. W. R. Snow,
      D.S.O., M.C. . . . Floringhem
Canadian 13 (R.E.8) Maj. A. G. R. Garrod,
      M.C. . . . . . Le Hameau
VIII   16 (R.E.8) Maj. A. W. C. V. Parr Camblain l'Abbé
XVII   52 (R.E.8) Maj. A. M. Morison Le Hameau
     'L' Flight (Bristol Fighter) Capt.
      B. E. Catchpole, M.C. . Bruay
     *Total aeroplanes on charge* . . 94 .

H.Q. *Tenth (Army) Wing.* Lieut.-Col. C. T.
  Maclean, M.C. . . . . . Sautrecourt
*Squadrons.*
   18 (D.H.4) Maj. G. R. Howard, D.S.O. Serny
   19 (Sopwith 'Dolphin') Maj. E. R. Prety-
    man . . . . . . Savy
   22 (Bristol Fighter) Maj. J. A. McKelvie Maisoncelle
   40 (S.E.5a) Maj. A. W. Keen, M.C. . Bryas
   64 (S.E.5a) Maj. B. E. Smythies . . Le Hameau
   148 (F.E.2b) Maj. I. T. Lloyd . . Sains-les-Pernes
   203 (Sopwith 'Camel') Maj. R. Collishaw,
    D.S.O., D.S.C., D.F.C. . . Filescamps Farm (Le
                                 Hameau)
   208 (Sopwith 'Camel') Maj. C. Draper,
    D.S.C. . . . . . Tramecourt
   *Total aeroplanes on charge* . . 156

## ORDER OF BATTLE

H.Q. *First Balloon Wing.* Lieut.-Col. P. K.
  Wise, D.S.O. . . . . . Bruay
  Companies 1 (Capt. W. Y. Walls, D.F.C., Sections 8 and 20)
           2 (Capt. C. M. Down, Sections 10 and 24)
           3 (Maj. G. T. J. Barry, Sections 21, 30, and 37)
           4 (Maj. J. R. Bedwell, M.C., Sections 42 and 46)
          10 (Capt. E. B. Cowell, Sections 5 and 28)
First Aircraft Park. Maj. H. W. Mills. . Lugy

H.Q. *II Brigade.* Brig.-Gen. T. I. Webb-Bowen,
  C.M.G., D.S.O. . . . . . St. Omer
H.Q. *Second (Corps) Wing.* Lieut.-Col. A. S.
  Barratt, M.C. . . . . . Oxelaere (Cassel)

| Corps. | Squadrons. | |
|---|---|---|
| XV | 4 (R.E.8) Maj. R. E. Saul . | St. Omer |
| II | 7 (R.E.8) Maj. B. E. Sutton, D.S.O., M.C. . . . . . | Droglandt |
| XIX | 10 (A.W.) Maj. K. D. P. Murray, M.C. . . . . . | Droglandt |
| X | 53 (R.E.8) Maj. G. Henderson . | Clairmarais |
|  | 82 (A.W.) Maj. J. B. Solomon, M.C. | Quelmes |
|  | *Total aeroplanes on charge* . . | 81 |

H.Q. *Eleventh (Army) Wing.* Lieut.-Col.
  H. A. Van Ryneveld, M.C. . . . Le Nieppe
  Squadrons.
  20 (Bristol Fighter) Maj. E. H. Johnston . Boisdinghem
  29 (S.E.5a) Maj. C. H. Dixon, M.C. . Hoog Huys
  70 (Sopwith 'Camel') Maj. E. L. Foot, M.C. Esquerdes
  74 (S.E.5a) Maj. K. L. Caldwell, M.C. . Clairmarais
  79 (Sopwith 'Dolphin') Maj. A. R. Arnold,
    D.F.C. . . . . . . St. Marie-Cappel
  206 (D.H.9) Maj. C. T. Maclaren . . Alquines
  149 (F.E.2b) Maj. B. P. Greenwood . Alquines
    (less 1 flight)
  *Total aeroplanes on charge* . . 153

H.Q. *Second Balloon Wing.* Lieut.-Col. H. M.
  Meyler, M.C. . . . . . Oxelaere (Cassel)
  Companies 5 (Capt. J. S. Giffard. Sections 2 and 25)
           6 (Capt. W. F. N. Forrest. Sections 9, 16, and 32)
           7 (Maj. F. X. Russell. Sections 15 and 38)
           8 (Maj. J. A. Cochrane. Sections 23 and 39)
          17 (Capt. G. F. M. Warner. Sections 13, 18, and 36)
Second Aircraft Park. Maj. C. G. Martyn . Houlle

## 8TH AUGUST 1918

H.Q. *III Brigade.* Brig.-Gen. C. A. H. Long-
croft, D.S.O. . . . . . Chateau Vaulx
H.Q. *Twelfth (Corps) Wing.* Lieut.-Col. A. B.
Burdett, D.S.O. . . . . Fienvillers

| Corps. | Squadrons. | |
|---|---|---|
| VI | 12 (R.E.8) Maj. H. S. Lees-Smith | Soncamp |
| V | 15 (R.E.8) Maj. H. V. Stammers . | Vert Galand |
| IV | 59 (R.E.8) Maj. C. J. Mackay, M.C. | Vert Galand |
| | *Total aeroplanes on charge* . . | 61 |

H.Q. *Thirteenth (Army) Wing.* Lieut.-Col.
P. H. L. Playfair, M.C. . . . Bachimont

*Squadrons.*
3 (Sopwith 'Camel') Maj. R. St. Clair
McClintock, M.C. . . . Valheureux
11 (Bristol Fighter) Maj. R. W. Heath . Le Quesnoy
56 (S.E.5a) Maj. E. J. L. W. Gilchrist,
M.C. . . . . . . Valheureux
57 (D.H.4) Maj. C. A. A. Hiatt, M.C. . Le Quesnoy
60 (S.E.5a) Maj. A. C. Clarke . . Boffles
87 (Sopwith 'Dolphin') Maj. C. J. W.
Darwin . . . . . Rougefay
102 (F.E.2b) Maj. F. C. Baker . . Surcamps
*Total aeroplanes on charge* . . 136

H.Q. *Third Balloon Wing.* Lieut.-Col. G. F. H.
Faithfull . . . . . . Fienvillers
*Companies* 12 (Capt. D. C. Bauer, D.F.C. Sections 35, 41, and 45)
18 (Lieut. D'A. J. Prendergast. Sections 1 and 31)
19 (Maj. W. S. Huxley, M.C. Sections 11 and 44)
Third Aircraft Park. Maj. O. V. Thomas . Candas

H.Q. *V. Brigade.* Brig.-Gen. L. E. O. Charlton,
C.M.G., D.S.O. . . . . Vauchelles-les-Domart
H.Q. *Fifteenth (Corps) Wing.* Lieut.-Col. J. A.
Chamier, O.B.E., D.S.O. . . . Vaux-en-Amienois

| Corps. | Squadrons. | |
|---|---|---|
| Australian | 3 (A.F.C.) (R.E.8) Maj. D. V. J. Blake . . . . | Villers Bocage |
| Canadian | 5 (R.E.8) Maj. C. H. Gardner | Bovelles |
| Cavalry | 6 (R.E.8) Maj. G. C. Pirie, M.C. | Bovelles |
| Tank | 8 (A.W.) Maj. T. L. Leigh-Mallory . . . . | Vignacourt |
| XXII | 9 (R.E.8) Maj. J. T. Rodwell . | Quevauvillers |
| III | 35 (A.W.) Maj. K. F. Balmain | Villers Bocage |
| | *Total aeroplanes on charge* . . | 110 |

# 120 ORDER OF BATTLE

H.Q. *Twenty-Second (Army) Wing.* Lieut.-Col. T. A. E. Cairnes, D.S.O. . . Bertangles

*Squadrons.*
  23 (Sopwith 'Dolphin') Maj. C. E. Bryant, D.S.O. . . . . . Bertangles
  24 (S.E.5a) Maj. V. A. H. Robeson, M.C. Conteville
  41 (S.E.5a) Maj. G. H. Bowman, D.S.O., M.C. . . . . . . Conteville
  48 (Bristol Fighter) Maj. K. R. Park, M.C. . . . . . . Bertangles
  65 (Sopwith 'Camel') Maj. H. V. Champion de Crespigny, M.C. . . . Bertangles
  80 (Sopwith 'Camel') Maj. V. D. Bell . Vignacourt
  84 (S.E.5a) Maj. W. S. Douglas, M.C. . Bertangles
  101 (F.E.2b) Maj. E. L. M. L. Gower . Famechon
  201 (Sopwith 'Camel') Maj. C. D. Booker, D.S.C. . . . . . . Poulainville
  205 (D.H.4) Maj. S. J. Goble, D.S.O., D.S.C. . . . . . Conteville (Bois de Roche)
  209 (Sopwith 'Camel') Maj. J. O. Andrews, D.S.O., M.C. . . . . Poulainville
  *Total aeroplanes on charge* . . 222

H.Q. *Fifth Balloon Wing.* Lieut.-Col. W. F. MacNeece, D.S.O. . . . . Querrieu
*Companies* 13 (Maj. G. S. Sansom, M.C. Sections 3 and 29)
          14 (Maj. H. P. L. Higman. Sections 14 and 19)
          15 (Capt. H. S. Goodliffe. Sections 6 and 22)
          16 (Capt. Hon. E. G. W. T. Knollys. Sections 12 and 43)
Fourth Aircraft Park. Maj. F. Jolly . . St. Riquier

H.Q. *X. Brigade.* Brig.-Gen. E. R. Ludlow-Hewitt, D.S.O., M.C.. . . . Wandonne
H.Q. *Sixty-Fifth Wing.* Lieut.-Col. J. A. Cunningham, D.F.C. . . . . Malo-les-Bains
*Squadrons.*
  17 (American) (Sopwith 'Camel') Lieut. E. B. Eckert . . . . Petite Synthe
  108 (D.H.9) Maj. S. S. Halse . . Cappelle
  148 (American) (Sopwith 'Camel') Lieut. M. L. Newhall . . . . Cappelle
  211 (D.H.9) Maj. G. R. M. Reid, M.C. Petite Synthe
  *Total aeroplanes on charge* . . 71

## 8TH AUGUST 1918

H.Q. *Eightieth (Army) Wing.* Lieut.-Col. L. A. Strange, M.C. . . . . Enguinegatte

*Squadrons.*
   2 (A.F.C.) (S.E.5a) Maj. A. Murray-Jones, M.C., D.F.C. . . . Reclinghem
   4 (A.F.C.) (Sopwith 'Camel') Maj. W. A. McClaughry, M.C. . . . Reclinghem
   46 (Sopwith 'Camel') Maj. A. H. O'Hara Wood. . . . . . Serny
   88 (Bristol Fighter) Maj. R. T. Leather . Serny
   92 (S.E.5a) Maj. A. Coningham, D.S.O., M.C. . . . . . . Serny
   103 (D.H.9) Maj. E. N. Fuller . . Serny
   *Total aeroplanes on charge* . . 126

H.Q. *Eighty-First (Corps) Wing.* Lieut.-Col. T. W. C. Carthew, D.S.O. . . . Enguinegatte

*Corps. Squadrons.*
  XIII  21 (R.E.8) Maj. L. T. N. Gould, M.C. . . . . . Floringhem
  XI  42 (R.E.8) Maj. H. J. F. Hunter, M.C. . . . . . Rely
   *Total aeroplanes on charge* . . 44

H.Q. *Eighth Balloon Wing.* Lieut.-Col. C. H. Stringer . . . . . Auchy-au-Bois
*Companies* 11 (Capt. T. G. G. Bolitho, M.C. Sections 40 and 47)
         20 (Capt. H. B. T. Hawkins. Sections 4, 34, and 48)
Tenth Aircraft Park. Maj. E. W. Havers . Ouve-Wirquin

H.Q. *Fifth (Operations) Group.* Brig.-Gen. C. L. Lambe, C.M.G., D.S.O. . . . Spycker
H.Q. *Sixty-First Wing.* Lieut.-Col. E. Osmond . . . . . St. Pol, Dunkirk

*Squadrons.*
   202 (D.H.4 and D.H.9) Maj. R. W. Gow, D.S.O., D.S.C. . . . . Bergues
   204 (Sopwith 'Camel') Maj. E. W. Norton, D.S.C. . . . . . Teteghem
   210 (Sopwith 'Camel') Maj. B. C. Bell, D.S.O., D.S.C. . . . . Eringhem
   213 (Sopwith 'Camel') Maj. R. Grahame, D.S.O., D.S.C. . . . . Bergues
   217 (D.H.4) Maj. W. L. Welsh, D.S.C. . Crochte
   *Total aeroplanes on charge* . . 88

# ORDER OF BATTLE

H.Q. *Eighty-Second Wing.* Lieut.-Col. C. F. Kilner, D.S.O. . . . . . Hames-Boucres
*Squadrons.*
  38 (F.E.2b) Maj. C. C. Wigram . . Cappelle
  214 (H.P.) Maj. H. G. Brackley, D.S.O., D.S.C. . . . . . St. Inglevert
  218 (D.H.9) Maj. B. S. Wemp, D.F.C. . Fretnum
    *Total aeroplanes on charge* . . 45
Eleventh Aircraft Park. Maj. J. B. Vernon . Coudekerque

INDEPENDENT FORCE.
  *General Officer Commanding.* Maj.-Gen. Sir Hugh M. Trenchard, K.C.B., D.S.O. . Autigny-la-Tour

H.Q. *VIII Brigade.* Brig.-Gen. C. L. N. Newall Froville
H.Q. *Forty-First Wing.* Lieut.-Col. J. E. A. Baldwin, D.S.O. . . . . Lupcourt
*Squadrons.*
  55 (D.H.4) Maj. A. Gray, M.C. . . Azelot
  99 (D.H.9) Maj. L. A. Pattinson, M.C.. Azelot
  104 (D.H.9) Maj. J. C. Quinnell . . Azelot
    *Total aeroplanes on charge* . . 58

H.Q. *Eighty-Third Wing.* Lieut.-Col. J. H. A. Landon, D.S.O. . . . . Bainville
*Squadrons.*
  100 (F.E.2b) Maj. C. G. Burge . . Ochey
  216 (H.P.) Maj. H. A. Buss, D.S.C. . Ochey
    *Total aeroplanes on charge* . . 28

Sixth Aircraft Park. Maj. A. F. A. Hooper, O.B.E. . . . . . . Vezelise

No. 1 Aircraft Depot. Lieut.-Col. S. A. Hebden . . . . . . Guines
No. 1 Aeroplane Supply Depot. Lieut.-Col. B. F. Moore . . . . . Marquise
No. 2 Aircraft Depot. Lieut.-Col. A. Christie, D.S.O. . . . . . . Groffliers
No. 2 Aeroplane Supply Depot. Lieut.-Col. V. Bettington . . . . . Berck-sur-Mer
No. 3 Aircraft Depot } Lieut.-Col.
No. 3 Aeroplane Supply } W. Wright, D.S.O. Courban
Depot

## 8TH AUGUST 1918

Sixth Aircraft Park (*cont.*):
No. 4 Aeroplane Supply Depot. Lieut.-Col.
    E. W. Stedman . . . . . Guines
Engine Repair Shops. Lieut.-Col. G. B.
    Hynes, D.S.O. . . . . . Pont de l'Arche, Rouen
    *Strength.* 93 Squadrons.
        2 Flights.
        1,782 Aeroplanes.
        43 K.B. Sections.

## APPENDIX XXV

### THE BATTLE OF AMIENS

*Memorandum by G.O.C., V Brigade, Royal Air Force, 14th August 1918*

O/C. 15th Wing, R.A.F.
O/C. 22nd Wing, R.A.F.

In addition to the special instructions issued by you on the subject of offence against Anti-Tank guns, I should like the following to be communicated to all pilots and observers participating in the battle when it is renewed:

1. All experience since the opening of the battle goes to prove the controlling action taken by the Anti-Tank guns of the enemy. Single guns have been responsible for 'knocking out' as many as 8 tanks in succession and thus completely holding up the advance in the sector concerned.

It is not too much to say that without the Anti-Tank gun the advance of our line would be irresistible.

The importance therefore of offensive action on the part of pilots and observers against these guns becomes of paramount importance and no opportunity should be missed; ground in front of the tank advance should be watched for their appearance and for their flashes, and it will be seldom that the duty in which machines are at the moment engaged will not yield in importance to offensive action at once against the Anti-Tank gun.

The above applies more especially to scouts engaged in low-flying attack, and also to Corps machines conducting Artillery, Contact, and Counter attack patrols according as the occasion may arise.

It is not possible to emphasize too strongly the duty and responsibility of pilots and observers in regard to the foregoing.

2. As is well known by now to all pilots and observers of Corps squadrons, their wireless calls will largely control the action of the low-flying scouts, subsequent to zero hour.

This system marks a large stride in the direction of close co-operation in battle between the Wings of a R.A.F. Brigade. The result will also inevitably tend towards economy of energy and greater efficiency in applying force at the right point of attack. To be successful, however, and attain the desired end, all such pilots and observers must be deeply imbued with their large responsibility in the matter and neglect no opportunity of furthering the success of the scheme.

3. It is to be carefully noted that at the outset of the battle, when renewed, the tanks will be in close support of the infantry, and not, as on the former occasion, leading them. This will require most accurate discrimination between friendly and hostile infantry in advance of the moving tank line.

4. Finally, as before, the R.A.F. action in the forthcoming battle should be largely decisive of its result, and the G.O.C., the Brigade, feels nothing with greater confidence than that the action of the pilots and observers who participate will be, if not surpassing, at any rate equal to their former effort on the 8th August and succeeding day.

*(Sgd.)* L. CHARLTON.
*Brigadier-General.*
*Commanding 5th Brigade, Royal Air Force.*

In the Field.
14th August 1918.

# APPENDIX XXVI
## STRENGTH OF ROYAL AIR FORCE, WESTERN FRONT
(INCLUDING INDEPENDENT FORCE AND 5th GROUP), 11th NOV. 1918

| Brigade. | Wing.* | No. of Squadron. | Aircraft. Type. | Duty.† |
|---|---|---|---|---|
| IX (H.Q.) | 9th (112) | H.Q. Commun. | *18 misc.* | |
| | | 25 | D.H.4 | D.B. |
| | | 27 | D.H.9 | D.B. |
| | | 32 | S.E.5a | S.S.F. |
| | | 49 | D.H.9 | D.B. |
| | | 62 | B.F. | F.R. |
| | | 18 | D.H.4 and D.H.9 (*a*) | D.B. |
| | 51st (89) | 1 | S.E.5a | S.S.F. |
| | | 43 | Sopwith 'Snipe' | S.S.F. |
| | | 94 | S.E.5a | S.S.F. |
| | | 107 | D.H.9 | D.B. |
| | | 205 | D.H.4 | D.B. |
| | 54th (51) | 83 | F.E.2b | N.B. |
| | | 151 | Sopwith 'Camel' | S.S.F. |
| | | 207 | H.P. | N.B. |
| | 82nd (43) | 58 | H.P. | N.B. |
| | | 152 | Sopwith 'Camel' | S.S.F. |
| | | 214 | H.P. | N.B. |
| TOTALS | 4 Wings (295) | | | |
| I | 1st (Corps) (64) | 5 | R.E.8 | C.R. |
| | | 16 | R.E.8 | C.R. |
| | | 52 | R.E.8 | C.R. |
| | | 'L' Flight | B.F. | F.R. |
| | 10th (Army) (95) | 98 | D.H.9 | D.B. |
| | | 22 | B.F. | F.R. |
| | | 40 | S.E.5a | S.S.F. |
| | | 148 | F.E.2b | N.B. |
| | | 203 | Sopwith 'Camel' | S.S.F. |
| | | 'I' Flight | F.E.2b | N.B. |
| | 91st (56) | 19 | Sopwith 'Dolphin' | S.S.F. |
| | | 64 | S.E.5a | S.S.F. |
| | | 209 | Sopwith 'Camel' | S.S.F. |
| TOTALS | 3 Wings (215) | | | |

* The figures in brackets represent the total number of aeroplanes under the Wing.
† D.B. = Day Bomber. S.S.F. = Single-Seater Fighter. F.R. = Fighter Reconnaissance. N.B. = Night Bomber. C.R. = Corps Reconnaissance.

# STRENGTH OF ROYAL AIR FORCE

| Brigade. | Wing.* | No. of Squadron. | Aircraft. | |
|---|---|---|---|---|
| | | | Type. | Duty.† |
| II | | H.Q. Commun. | 18 misc. | |
| | 2nd (Corps) (98) | 4 | R.E.8 | C.R. |
| | | 7 | R.E.8 | C.R. |
| | | 10 | A.W.B. | C.R. |
| | | 53 | R.E.8 | C.R. |
| | | 82 | A.W.B. | C.R. |
| | | 'M' Flight | B.F. | F.R. |
| | 11th (Army) (151) | 29 | S.E.5a | S.S.F. |
| | | 41 | S.E.5a | S.S.F. |
| | | 48 | B.F. | F.R. |
| | | 70 | Sopwith 'Camel' | S.S.F. |
| | | 74 | S.E.5a | S.S.F. |
| | | 79 | Sopwith 'Dolphin' | S.S.F. |
| | | 149 | F.E.2b | N.B. |
| | | 206 | D.H.9 | D.B. |
| | 65th (85) | 24 | S.E.5a | S.S.F. |
| | | 38 | F.E.2b | N.B. |
| | | 65 | Sopwith 'Camel' | S.S.F. |
| | | 108 | D.H.9 | D.B. |
| | | 204 | Sopwith 'Camel' | S.S.F. |
| | | 39 | B.F. | F.R. |
| Totals | 3 Wings (334) | | | |
| III | 12th (Corps) (76) | 12 | R.E.8 | C.R. |
| | | 13 | R.E.8 | C.R. |
| | | 15 | R.E.8 | C.R. |
| | | 59 | R.E.8 | C.R. |
| | 13th (Army) (94) | 'N' Flight | B.F. | F.R. |
| | | 56 | S.E.5a | S.S.F. |
| | | 60 | S.E.5a | S.S.F. |
| | | 87 | Sopwith 'Dolphin' | S.S.F. |
| | | 201 | Sopwith 'Camel' | S.S.F. |
| | | 210 | Sopwith 'Camel' | S.S.F. |
| | 90th (78) | 3 | Sopwith 'Camel' | S.S.F. |
| | | 11 | B.F. | F.R. |
| | | 57 | D.H.4 | D.B. |
| | | 102 | F.E.2b | N.B. |
| Totals | 3 Wings (248) | | | |

# WESTERN FRONT

| Brigade. | Wing.* | No. of Squadrons. | Aircraft. Type. | Duty.† |
|---|---|---|---|---|
| V | 15th (Corps) (86) | H.Q. (Commun.) 3 (A.F.C.) 8 9 35 73 | 18 misc. R.E.8 A.W.B. R.E.8 A.W.B. Sopwith 'Camel' | C.R. C.R. C.R. C.R. S.S.F. |
| | 22nd (Army) (181) | 'O' Flight 46 80 84 208 20 23 92 211 218 | B.F. Sopwith 'Camel' Sopwith 'Camel' S.E.5a Sopwith 'Camel' B.F. Sopwith 'Dolphin' S.E.5a D.H.9 D.H.9 | F.R. S.S.F. S.S.F. S.S.F. S.S.F. F.R. S.S.F. S.S.F. D.B. D.B. |
| | —(18) | 101 | F.E.2b | N.B. |
| TOTALS | 2 Wings (285) | | | |
| X | 81st (Corps) (81) | 2 21 42 6 | A.W.B. R.E.8 R.E.8 R.E.8 | C.R. C.R. C.R. C.R. |
| | 80th (Army) (120) | 'P' Flight 2 (A.F.C.) 4 (A.F.C.) 88 103 54 85 | B.F. S.E.5a Sopwith 'Snipe' B.F. D.H.9 Sopwith 'Camel' S.E.5a | F.R. S.S.F. S.S.F. F.R. D.B. S.S.F. S.S.F. |
| TOTALS | 2 Wings (201) | | | |

\* The figures in brackets represent the total number of aeroplanes under the Wing.
† D.B.= Day Bomber.  S.S.F.= Single-Seater Fighter.  F.R.= Fighter Reconnaissance.  N.B. = Night Bomber.  C.R. = Corps Reconnaissance.

# STRENGTH OF ROYAL AIR FORCE

| Brigade. | Wing.* | No. of Squadron. | Aircraft. | |
|---|---|---|---|---|
| | | | Type. | Duty.† |
| | | H.Q. Commun. | 18 misc. | |
| | *Independent Force.* | | | |
| VIII | 41st Wing (59) | 55 | D.H.4 | D.B. |
| | | 99 | D.H.9 (a) | D.B. |
| | | 104 | D.H.9 | D.B. |
| | 83rd Wing (49) | 97 | H.P. | N.B. |
| | | 100 | H.P. | N.B. |
| | | 115 | H.P. | N.B. 4 |
| | | 215 | H.P. | N.B. |
| | | 216 | H.P. | N.B. |
| | 88th Wing (32) | 45 | Sopwith 'Camel' | S.S.F. |
| | | 110 | D.H.9 (a) | D.B. |
| TOTALS | 3 Wings (140) | | | |
| | *5th Group.* | | | |
| | | 202 | D.H.4 | D.B. |
| | | 213 | Sopwith 'Camel' | S.S.F. |
| | | 217 | D.H.4 | D.B. |
| | (63) | Flight No. 471 | Sopwith 'Camel' | S.S.F. |

\* The figures in brackets represent the total number of aeroplanes under the Wing.
† D.B. = Day Bomber. S.S.F. = Singer Seater Fighter. F.R. = Fighter Reconnaissance. N.B. = Night Bomber. C.R. = Corps Reconnaissance.

## SUMMARY

### Western Front

| Unit. | No. of Wings. | No. of Squadrons. | No. of Flights. | Aircraft. | | Total. |
|---|---|---|---|---|---|---|
| | | | | Serviceable. | Unserviceable. | |
| G.H.Q. | .. | 1 | .. | 18 | .. | 18 |
| IX H.Q. Bde. | 4 | 17 | .. | 248 | 47 | 295 |
| I Bde. | 3 | 11 | 2 | 195 | 20 | 215 |
| II Bde. | 3 | 19 | 1 | 295 | 39 | 334 |
| III Bde. | 3 | 13 | 1 | 219 | 29 | 248 |
| V Bde. | 2 | 15 | 1 | 233 | 52 | 285 |
| X Bde. | 2 | 10 | 1 | 180 | 21 | 201 |
| I.F., VIII Bde. | 3 | 10 | .. | 127 | 13 | 140 |
| 5th Group. | .. | 3 | 1 | 61 | 2 | 63 |
| TOTALS | 20 | 99 | 7 | 1,576 | 223 | 1,799 |

# WESTERN FRONT

*Detail of Employment of Squadrons and Flights*

| Unit. | Commun. Sqdn. | Corps Reconn. | Fighter Reconn. | Day Bomb- ing. | Short Night Bombing. | Long Night Bomb- ing. | Single- seater Fighting. | Total. |
|---|---|---|---|---|---|---|---|---|
| G.H.Q. | 1 | .. | .. | .. | .. | .. | .. | 1 |
| IX H.Q. Bde. | .. | .. | 1 | 6 | 1 | 3 | 6 | 17 |
| I Bde. | .. | 3 | 1 (plus 1 Flt.) | 1 | 1 (plus 1 Flt.) | .. | 5 | 11 (plus 2 Flts.) |
| II Bde. | .. | 5 | 2 (plus 1 Flt.) | 2 | 2 | .. | 8 | 19 (plus 1 Flt.) |
| III Bde. | .. | 4 | 1 (plus 1 Flt.) | 1 | 1 | .. | 6 | 13 (plus 1 Flt.) |
| V Bde. | .. | 4 | 1 (plus 1 Flt.) | 2 | 1 | .. | 7 | 15 (plus 1 Flt.) |
| X Bde. | .. | 4 | 1 (plus 1 Flt.) | 1 | .. | .. | 4 | 10 (plus 1 Flt.) |
| I.F., VIII Bde. | .. | .. | .. | 4 | .. | 5 | 1 | 10 |
| 5th Group | .. | .. | .. | 2 | .. | .. | 1 (plus 1 Flt.) | 3 (plus 1 Flt.) |
| TOTAL | 1 | 20 | 7 (plus 5 Flts.) | 19 | 6 (plus 1 Flt.) | 8 | 38 (plus 1 Flt.) | 99 (plus 7 Flts.) |

*Detail of Aircraft employed*

| Type of duty. | Approx. no. engaged. | Remarks. |
|---|---|---|
| Corps Reconnaissance | 370 | 20 Sqdns. at average 18·5 |
| Fighter Reconnaissance | 138 | 6 Sqdns. at average 18 (excluding 39 Sqdn.), 5 Flts. at 6. |
| Day Bombing | 333 | 19 Sqdns. at average 17·5. |
| Short Night Bombing | 113 | 6 Sqdns. at average 18·5 plus 1 Flt. of 2 m/cs. |
| Long Night Bombing. | 80 | 8 Sqdns. at 10. |
| Single-seater Fighting. | 747 | 38 Sqdns. at 19·5 plus 1 Flt. at 6. |
| G.H.Q. Commun. duty | 18 | 1 Sqdn. at 18. |
| TOTAL | 1,799 | |

# APPENDIX XXVIII

## LIST OF SQUADRONS R.F.C. AND R.A.F. WHICH SERVED ON THE WESTERN FRONT 1914–18

| Squadron No. | Date of arrival in France. | Duties. | Aeroplanes. | |
|---|---|---|---|---|
| | | | Type. | Engine. |
| 1 | 7.3.15 | General (Mar. 1915– Feb. 1916) | Avro | 80-h.p. Gnome |
| | | | B.E.8 | ,, ,, |
| | | | Caudron | ,, ,, |
| | | | Martinsyde Scout | ,, ,, |
| | | | Morane 'Parasol' | 80-h.p. Le Rhône |
| | | | Bristol Scout | ,, ,, |
| | | | Morane Biplane | 110-h.p. Le Rhône |
| | | Fighting (Feb. 1916– Nov. 1918) | Nieuport Scout | ,, ,, |
| | | | Nieuport Two-seater | 110-h.p. Clerget |
| | | | S.E.5a | 200-h.p. Wolseley 'Viper' |
| 2 | 13.8.14 | General (Aug. 1914– Feb. 1916) | B.E.2a | 70-h.p. Renault |
| | | | B.E.2b | ,, ,, |
| | | | B.E.2c | ,, ,, |
| | | | R.E.1 | ,, ,, |
| | | | R.E.5 | 120-h.p. Beardmore |
| | | | Vickers Fighter | 100-h.p. Gnome Monosoupape |
| | | | Maurice Farman | 80-h.p. Renault |
| | | Corps (Feb. 1916– Nov. 1918) | B.E.2c | 90-h.p. R.A.F. 1a |
| | | | B.E.2d | ,, ,, |
| | | | B.E.2e | ,, ,, |
| | | | A.W. | 160-h.p. Beardmore |
| 3 | 13.8.14 | General (Aug. 1914– Feb. 1916) | Bleriot | 80-h.p. Gnome |
| | | | Henri Farman | ,, ,, |
| | | | B.E.8 | ,, ,, |
| | | | Sopwith Scout ('Tabloid') | ,, ,, |
| | | | Bristol Scout | ,, ,, |
| | | | S.E.2 | ,, ,, |
| | | | Avro | ,, ,, |
| | | Corps (Feb. 1916– Oct. 1917) | Morane 'Parasol' | 80-h.p. Le Rhône |
| | | | Morane Biplane | 110-h.p. Le Rhône |
| | | Fighting (Oct. 1917– Nov. 1918) | Sopwith Camel | ,, ,, |
| 4 'A' and 'B' Flights | 13.8.14 | General (Aug. 1914– Feb. 1916) | B.E.2a | 70-h.p. Renault |
| | | | B.E.2b | ,, ,, |
| | | | B.E.2c | ,, ,, |

# SQUADRONS ON WESTERN FRONT 1914–18

| Squadron No. | Date of arrival in France. | Duties. | Aeroplanes. | |
|---|---|---|---|---|
| | | | Type. | Engine. |
| 'C' Flight | 20.9.14 | General (Aug. 1914– Feb. 1916) | B.E.2c | 90-h.p. R.A.F. 1a |
| | | | Maurice Farman | 70-h.p. Renault |
| | | | Sopwith Scout | 80-h.p. Gnome |
| | | | Voisin | 130-h.p. Canton-Unne |
| | | | Martinsyde Scout | 80-h.p. Gnome |
| | | | Bristol Scout | ,, ,, |
| | | | Caudron | ,, ,, |
| | | | Morane Monoplane | ,, ,, |
| | | Corps (Feb. 1916– Nov. 1918) | B.E.2c | 90-h.p. R.A.F. 1a |
| | | | B.E.2d | ,, ,, |
| | | | B.E.2e | ,, ,, |
| | | | R.E.8 | 150-h.p. R.A.F. 4a |
| 5 | 15.8.14 | General (Aug. 1914– Feb. 1916) | Henri Farman | 80-h.p. Gnome |
| | | | Avro | ,, ,, |
| | | | B.E.8 | ,, ,, |
| | | | Bristol Scout | ,, ,, |
| | | | Martinsyde Scout | ,, ,, |
| | | | Voisin | 130-h.p. Canton-Unne |
| | | | Vickers Fighter | 100-h.p. Gnome Monosoupape |
| | | | Bleriot | 80-h.p. Gnome |
| | | | Caudron | ,, ,, |
| | | | D.H.2 | 100-h.p. Gnome Monosoupape |
| | | | F.E.8 | 100-h.p. Gnome Monosoupape |
| | | Corps (Feb. 1916– Nov. 1918) | B.E.2c | 90-h.p. R.A.F. 1a |
| | | | B.E.2d | ,, ,, |
| | | | B.E.2e | ,, ,, |
| | | | R.E.8 | 150-h.p. R.A.F. 4a |
| 6 | 7.10.14 | General (Oct. 1914– Feb. 1916) | B.E.2a | 70-h.p. Renault |
| | | | B.E.2b | ,, ,, |
| | | | B.E.2c | ,, ,, |
| | | | Henri Farman | 80-h.p. Gnome |
| | | | Bleriot | ,, ,, |
| | | | B.E.8 | ,, ,, |
| | | | Martinsyde Scout | 80-h.p. Gnome / 120-h.p. Beardmore |
| | | | F.E.2a | 120-h.p. Beardmore |
| | | | F.E.2b | ,, ,, |
| | | | Bristol Scout | 80-h.p. Gnome |
| | | Corps (Feb. 1916– Nov. 1918) | B.E.2c | 90-h.p. R.A.F. 1a |
| | | | B.E.2d | ,, ,, |
| | | | B.E.2e | ,, ,, |
| | | | R.E.8 | 150-h.p. R.A.F. 4a |

# LIST OF SQUADRONS

| Squadron No. | Date of arrival in France. | Duties. | Aeroplanes. Type. | Engine. |
|---|---|---|---|---|
| 7 | 8.4.15 | General (Apr. 1915–Feb. 1916) | Vickers Fighter | 100-h.p. Gnome Monosoupape |
| | | | R.E.5 | 120-h.p. Beardmore |
| | | | Voisin | 130-h.p. Canton-Unne |
| | | | B.E.2c | { 70-h.p. Renault <br> { 90-h.p. R.A.F. 1a |
| | | Corps (Feb. 1916–Nov. 1918) | B.E.2d | 90-h.p. R.A.F. 1a |
| | | | B.E.2e | ,, ,, |
| | | | Bristol Scout | 80-h.p. Le Rhône |
| | | | R.E.8 | 150-h.p. R.A.F. 4a |
| 8 | 15.4.15 | General (Apr. 1915–Feb. 1916) | B.E.2c | 90-h.p. R.A.F. 1a |
| | | | B.E.2d | ,, ,, |
| | | | B.E.2e | ,, ,, |
| | | Corps (Feb. 1916–Nov. 1918) | Bristol Scout | { 80-h.p. Gnome <br> { 80-h.p. Le Rhône |
| | | | A.W. | 160-h.p. Beardmore |
| 9* | 8.12.14 | Wireless (Dec. 1914–Mar. 1915) | B.E.2a | 70-h.p. Renault |
| | | | Maurice Farman | ,, ,, |
| | | | Bleriot | 80-h.p. Gnome |
| 9 | Rejoined B.E.F. 20.12.15 | General (Dec. 1915–Feb. 1916) | B.E.2c | 90-h.p. R.A.F. 1a |
| | | | B.E.2d | ,, ,, |
| | | | B.E.2e | ,, ,, |
| | | | Bristol Scout | 80-h.p. Le Rhône |
| | | Corps (Feb. 1916–Nov. 1918) | R.E.8 | 150-h.p. R.A.F. 4a |
| 10 | 25.7.15 | General (July 1915–Feb. 1916) | B.E.2c | 70-h.p. Renault |
| | | | ,, | 90-h.p. R.A.F. 1a |
| | | | B.E.2d | 90-h.p. R.A.F. 1a |
| | | | B.E.2e | ,, ,, |
| | | Corps (Feb. 1916–Nov. 1918) | Bristol Scout | 80-h.p. Le Rhône |
| | | | A.W. | 160-h.p. Beardmore |
| 11 | 25.7.15 | Fighting Reconnaissance (July 1915–Nov. 1918) | Vickers Fighter | 100-h.p. Gnome Monosoupape |
| | | | Bristol Scout | 80-h.p. Le Rhône |
| | | | D.H.2 | 100-h.p. Gnome Monosoupape |
| | | | Nieuport Scout | 110-h.p. Le Rhône |
| | | | Vickers Scout | 100-h.p. Le Rhône |
| | | | F.E.2b | { 120-h.p. Beardmore <br> { 160-h.p. Beardmore |
| | | | Bristol Fighter | 190–250-h.p. Rolls Royce 'Falcon' |

\* Formed at H.Q. St. Omer from the Head-quarter Wireless Unit as No. 9 (Wireless) Squadron. Disbanded in France 22.3.15. Re-formed in England 1.4.15.

ON WESTERN FRONT 1914-18

| Squadron No. | Date of arrival in France. | Duties. | Aeroplanes. | |
|---|---|---|---|---|
| | | | Type. | Engine. |
| 12 | 6.9.15 | General (Sept. 1915– Feb. 1916) | B.E.2c<br>B.E.2d<br>B.E.2e | 90-h.p. R.A.F. 1a<br>,, ,,<br>,, ,, |
| | | Corps (Feb. 1916– Nov. 1918) | R.E.8 | 150-h.p. R.A.F. 4a |
| 13 | 19.10.15 | General (Oct. 1915– Feb. 1916) | B.E.2c<br>B.E.2d<br>B.E.2e<br>Bristol Scout | 90-h.p. R.A.F. 1a<br>,, ,,<br>,, ,,<br>80-h.p. Le Rhône |
| | | Corps (Feb. 1916– Nov. 1918) | R.E.8 | 150-h.p. R.A.F. 4a |
| 15 | 23.12.15 | General (Jan. 1916– Feb. 1916) | B.E.2c<br>B.E.2d<br>B.E.2e<br>Bristol Scout | 90-h.p. R.A.F. 1a<br>,, ,,<br>,, ,,<br>80-h.p. Le Rhône |
| | | Corps (Feb. 1916– Nov. 1918) | R.E.8 | 150-h.p. R.A.F. 4a |
| 16 | Formed in France 10.2.15 | General (Feb. 1915– Feb. 1916) | B.E.2c<br>Maurice Farman<br>Voisin<br><br>Bleriot<br>R.E.5<br>Vickers Fighter<br><br>Martinsyde Scout<br>Bristol Scout<br>F.E.2b | 90-h.p. R.A.F. 1a<br>70-h.p. Renault<br>140-h.p. Canton-Unne<br>80-h.p. Gnome<br>120-h.p. Beardmore<br>100-h.p. Gnome Monosoupape<br>80-h.p. Gnome<br>,, ,,<br>120-h.p. Beardmore |
| | | Corps (Feb. 1916– Nov. 1918) | B.E.2c<br>B.E.2d<br>B.E.2e<br>R.E.8 | 90-h.p. R.A.F. 1a<br>,, ,,<br>,, ,,<br>150-h.p. R.A.F. 4a |
| 18 | 19.11.15 | Fighting Reconnaissance (Nov. 1915– June 1917) | Vickers Fighter<br><br>D.H.2<br><br>Bristol Scout<br>F.E.2b<br><br>Martinsyde Scout | 100-h.p. Gnome Monosoupape<br>100-h.p. Gnome Monosoupape<br>80-h.p. Le Rhône<br>{ 120-h.p. Beardmore<br>{ 160-h.p. Beardmore<br>{ 120-h.p. Beardmore<br>{ 160-h.p. Beardmore |
| | | Bombing (June 1917– Nov. 1918) | D.H.4<br>D.H.9a | 200-h.p. R.A.F. 3a<br>400-h.p. Liberty |

# LIST OF SQUADRONS

| Squadron No. | Date of arrival in France. | Duties. | Aeroplanes. | |
|---|---|---|---|---|
| | | | Type. | Engine. |
| 19 | 30.7.16 | Fighting (July 1916–Nov. 1918) | B.E.12 | 150-h.p. R.A.F. 4a |
| | | | Spad | { 150-h.p. Hispano-Suiza<br>200-h.p. Hispano-Suiza |
| | | | Sopwith 'Dolphin' | 200-h.p. Hispano-Suiza |
| 20 | 23.1.16 | Fighting Reconnaissance (Jan. 1916–Nov. 1918) | F.E.2b | 160-h.p. Beardmore |
| | | | Martinsyde Scout | { 120-h.p. Beardmore<br>160-h.p. Beardmore |
| | | | F.E.2d | 250-h.p. Rolls-Royce 'Eagle' |
| | | | Bristol Fighter | 190–250 h.p. Rolls-Royce 'Falcon' |
| 21 | 23.1.16 | Corps (Jan. 1916–Aug. 1916) | R.E.7 | 150-h.p. R.A.F. 4a |
| | | | B.E.2c | 90-h.p. R.A.F. 1a |
| | | | Martinsyde Scout | { 120-h.p. Beardmore<br>160-h.p. Beardmore |
| | | | Bristol Scout | 80-h.p. Le Rhône |
| | | Fighting (Aug. 1916–Feb. 1917) | B.E.12 | 150-h.p. R.A.F. 4a |
| | | Corps (Feb. 1917–Nov. 1918) | R.E.8 | 150-h.p. R.A.F. 4a |
| 22 | 1.4.16 | Fighting Reconnaissance (Apr. 1916–Nov. 1918) | F.E.2b | { 120-h.p. Beardmore<br>160-h.p. Beardmore |
| | | | Bristol Fighter | 190–250-h.p. Rolls-Royce 'Falcon' |
| 23 | 16.3.16 | Fighting Reconnaissance (Mar. 1916–Feb. 1917) | F.E.2b | { 120-h.p. Beardmore<br>160-h.p. Beardmore |
| | | | Martinsyde Scout | { 120-h.p. Beardmore<br>160-h.p. Beardmore |
| | | Fighting (Feb. 1917–Nov. 1918) | Spad | { 150-h.p. Hispano-Suiza<br>200-h.p. Hispano-Suiza |
| | | | Sopwith 'Dolphin' | 200-h.p. Hispano-Suiza |
| 24 | 7.2.16 | Fighting (Feb. 1916–Nov. 1918) | D.H.2 | 100-h.p. Gnome Monosoupape |
| | | | D.H.5 | 110-h.p. Le Rhône |
| | | | S.E.5 | 150-h.p. Hispano-Suiza |
| | | | S.E.5a | { 200-h.p. Hispano-Suiza<br>200-h.p. Wolseley 'Viper' |

## ON WESTERN FRONT 1914–18

| Squadron No. | Date of arrival in France. | Duties. | Aeroplanes. | |
|---|---|---|---|---|
| | | | Type. | Engine. |
| 25 | 20.2.16 | Fighting Reconnaissance (Feb. 1916–Nov. 1918) | F.E.2b<br>Bristol Scout<br>F.E.2d<br>D.H.4 | 120-h.p. Beardmore<br>160-h.p. Beardmore<br>80-h.p. Le Rhône<br>250-h.p. Rolls-Royce 'Eagle'<br>250-h.p. Rolls-Royce 'Eagle' |
| 27 | 1.3.16 | Fighting (Mar. 1916–June 1916)<br><br>Bombing (July 1916–Nov. 1918) | Martinsyde Scout<br><br>D.H.4<br><br>D.H.9 | 160-h.p. Beardmore<br><br>200-h.p. B.H.P.<br>200-h.p. Siddeley 'Puma'<br>200-h.p. B.H.P.<br>200-h.p. Siddeley 'Puma' |
| 28 | 8.10.17<br>To Italian Front<br>12.11.17 | Fighting (Oct. 1917–Nov. 1917) | Sopwith 'Camel' | 130-h.p. Clerget |
| 29 | 25.3.16 | Fighting (Mar. 1916–Nov. 1918) | D.H.2<br>Nieuport Scout<br><br>S.E.5a | 100-h.p. Gnome Monosoupape<br>110-h.p. Le Rhône<br>200-h.p. Hispano-Suiza<br>200-h.p. Wolseley 'Viper' |
| 32 | 28.5.16 | Fighting (May 1916–Nov. 1918) | D.H.2<br>D.H.5<br><br>S.E.5a | 100-h.p. Gnome Monosoupape<br>110-h.p. Le Rhône<br>200-h.p. Hispano-Suiza<br>200-h.p. Wolseley 'Viper' |
| 34 | 15.7.16<br>To Italian Front<br>13.11.17 | Corps (July 1916–Nov. 1917) | B.E.2e<br>R.E.8 | 90-h.p. R.A.F. 1a<br>150-h.p. R.A.F. 4a |
| 35 | 26.1.17 | Corps (Jan. 1917–Nov. 1918) | A.W. | 160-h.p. Beardmore |
| 38 | 31.5.18 | Night Bombing (May 1918–Nov. 1918) | F.E.2b | 160-h.p. Beardmore |
| 40 'A' Flight | 2.8.16 | Fighting (Aug. 1916–Nov. 1918) | F.E.8 | 100-h.p. Gnome Monosoupape |

# LIST OF SQUADRONS

| Squadron No. | Date of arrival in France. | Duties. | Aeroplanes. Type. | Aeroplanes. Engine. |
|---|---|---|---|---|
| 'B' and 'C' Flights | 25.8.16 | Fighting | Nieuport Scout | 110-h.p. Le Rhône |
| | | | S.E.5 | 150-h.p. Hispano-Suiza |
| | | | S.E.5a | 200-h.p. Hispano-Suiza |
| 41 | 15.10.16 | Fighting (Oct. 1916– Nov. 1918) | F.E.8 | 100-h.p. Gnome Monosoupape |
| | | | D.H.5 | 110-h.p. Le Rhône |
| | | | S.E.5a | 200-h.p. Hispano-Suiza<br>200-h.p. Wolseley 'Viper' |
| 42 | 8.8.16<br>To Italian Front Dec.1917.<br>Returned to France Mar. 1918 | Corps (Aug. 1916– Nov. 1918) | B.E.2d<br>B.E.2e<br>R.E.8 | 90-h.p. R.A.F. 1a<br>,,   ,,<br>150-h.p. R.A.F. 4a |
| 43 | 17.1.17 | Fighting Reconnaissance (Jan. 1917– Sept. 1917) | Sopwith Two-Seater ('1½ strutter') | 110-h.p. Clerget<br>130-h.p. Clerget |
| | | Fighting (Sept. 1917– Nov. 1918) | Sopwith 'Camel'<br>Sopwith 'Snipe' | 130-h.p. Clerget<br>200-h.p. B.R.2 |
| 45 | 12.10.16<br>To Italian Front Dec.1917.<br>To independent Air Force Sept.1918 | Fighting Reconnaissance (Oct. 1916– July 1917) | Sopwith Two-seater ('1½ strutter') | 110-h.p. Clerget<br>130-h.p. Clerget |
| | | Fighting (July 1917– Dec. 1917) | Sopwith 'Camel' | 130-h.p. Clerget |
| 46 | 20.10.16<br>To England for air raid duty 10.7.17.<br>Returned to France 30.8.17. | Corps (Oct. 1916– Apr. 1917) | Nieuport Two-seater | 110-h.p. Clerget |
| | | Fighting (Apr. 1917– Nov. 1918) | Sopwith Scout ('Pup')<br>Sopwith 'Camel' | 80-h.p. Le Rhône<br>110-h.p. Le Rhône |
| 48 | 8.3.17 | Fighting Reconnaissance (Mar. 1917– Nov. 1918) | Bristol Fighter | 190-h.p. Rolls-Royce 'Falcon' |

# ON WESTERN FRONT 1914–18

| Squadron No. | Date of arrival in France. | Duties. | Aeroplanes. | |
|---|---|---|---|---|
| | | | Type. | Engine. |
| 49 | 12.11.17 | Bombing (Nov. 1917–Nov. 1918) | D.H.4<br>D.H.9 | 200-h.p. R.A.F. 3a<br>230-h.p. B.H.P. |
| 52 | 21.11.16 | Corps (Nov. 1916–Nov. 1918) | B.E.2e<br>R.E.8 | 90-h.p. R.A.F. 1a<br>150-h.p. R.A.F. 4a |
| 53 | 1.1.17 | Corps (Jan. 1917–Nov. 1918) | B.E.2e<br>R.E.8 | 90-h.p. R.A.F. 1a<br>150-h.p. R.A.F. 4a |
| 54 | 24.12.16 | Fighting (Dec. 1916–Nov. 1918) | Sopwith Scout ('Pup')<br>Sopwith 'Camel' | 80-h.p. Le Rhône<br>{110-h.p. Le Rhône<br>{130-h.p. Clerget |
| 55 | 6.3.17<br>To 41st Wing 11.10.17. | Bombing (Mar. 1917–Nov. 1918) | D.H.4 | 250-h.p. Rolls-Royce 'Eagle' |
| 56 | 8.4.17 | Fighting (Apr. 1917–Nov. 1918) | S.E.5<br><br>S.E.5a | 150-h.p. Hispano-Suiza<br>{200-h.p. Hispano-Suiza<br>{200-h.p. Wolseley 'Viper' |
| 57 | 16.12.16 | Fighting Reconnaissance (Dec. 1916–May 1917)<br>Bombing (May 1917–Nov. 1918) | F.E.2d<br><br><br>D.H.4 | 250-h.p. Rolls-Royce 'Eagle'<br><br><br>250-h.p. Rolls-Royce 'Eagle' |
| 58 | 10.1.18 | Night Bombing (Jan. 1918–Nov. 1918) | F.E.2b<br>Handley Page | 160-h.p. Beardmore<br>2-275-h.p. Rolls-Royce 'Eagle' |
| 59 | 23.2.17 | Corps (Feb. 1917–Nov. 1918) | R.E.8 | 150-h.p. R.A.F. 4a |
| 60 | 28.5.16 | Fighting (May 1916–Nov. 1918) | Morane Scout<br>Morane Biplane<br>Morane 'Parasol'<br>Nieuport Scout<br>S.E.5 | {80-h.p. Le Rhône<br>{110-h.p. Le Rhône<br>{80-h.p. Le Rhône<br>{110-h.p. Le Rhône<br>80-h.p. Le Rhône<br>110-h.p. Le Rhône<br>150-h.p. Hispano-Suiza |

## LIST OF SQUADRONS

| Squadron No. | Date of arrival in France. | Duties. | Aeroplanes. | |
|---|---|---|---|---|
| | | | Type. | Engine. |
| | | Fighting | S.E.5a | 200-h.p. Hispano-Suiza<br>200-h.p. Wolseley 'Viper' |
| 62 | 23.1.18 | Fighting Reconnaissance (Jan. 1918–Nov. 1918) | Bristol Fighter | 190–250-h.p. Rolls-Royce 'Falcon' |
| 64 | 15.10.17 | Fighting (Oct. 1917–Nov. 1918) | D.H.5 | 110-h.p. Le Rhône |
| | | | S.E.5a | 200-h.p. Hispano-Suiza<br>200-h.p. Wolseley 'Viper' |
| 65 | 27.10.17 | Fighting (Oct. 1917–Nov. 1918) | Sopwith 'Camel' | 130-h.p. Clerget |
| 66 | 17.3.17<br>To Italian Front 22.11.17. | Fighting (Mar. 1917–Nov. 1917) | Sopwith Scout ('Pup') | 80-h.p. Le Rhône |
| 68 A.F.C. became No. 2 (A.F.C.) Squadron on 19.1.18. | 21.9.17 | Fighting (Sept. 1917–Nov. 1918) | D.H.5 | 110-h.p. Le Rhône |
| | | | S.E.5a | 200-h.p. Hispano-Suiza<br>200-h.p. Wolseley 'Viper' |
| 69 A.F.C. became No. 3 (A.F.C.) Squadron on 19.1.18. | 9.9.17 | Corps (Sept. 1917–Nov. 1918) | R.E.8 | 150-h.p. R.A.F. 4a |
| 70<br>'A' Flight<br>'B' Flight<br>'C' Flight | 24.5.16<br>29.6.16<br>30.7.16 | Fighting Reconnaissance (May 1916–July 1917) | Sopwith Two-seater ('1½ strutter') | 110-h.p. Clerget<br>130-h.p. Clerget |
| | | Fighting (July 1917–Nov. 1918) | Sopwith 'Camel' | 130-h.p. Clerget |

| Squadron No. | Date of arrival in France. | Duties. | Aeroplanes. | |
|---|---|---|---|---|
| | | | Type. | Engine. |
| 71 A.F.C. became No. 4 (A.F.C.) Squadron on 19.1.18. | 18.12.17 | Fighting (Dec. 1917– Nov. 1918) | Sopwith 'Camel' Sopwith 'Snipe' | 110-h.p. Le Rhône 130-h.p. Clerget 230-h.p. B.R.2 |
| 73 | 9.1.18 | Fighting (Jan. 1918– Nov. 1918) | Sopwith 'Camel' | 110-h.p. Le Rhône 130-h.p. Clerget |
| 74 | 30.3.18 | Fighting (Mar. 1918– Nov. 1918) | S.E.5a | 200-h.p. Wolseley 'Viper' |
| 79 | 22.2.18 | Fighting (Feb. 1918– Nov. 1918) | Sopwith 'Dolphin' | 200-h.p. Hispano-Suiza |
| 80 | 27.1.18 | Fighting (Jan. 1918– Nov. 1918) | Sopwith 'Camel' | 110-h.p. Le Rhône |
| 82 | 20.11.17 | Corps (Nov. 1917– Nov. 1918) | A.W. | 160-h.p. Beardmore |
| 83 | 4.3.18 | Night Bombing (Mar. 1918– Nov. 1918) | F.E.2b | 160-h.p. Beardmore |
| 84 | 23.9.17 | Fighting (Sept. 1917– Nov. 1918) | S.E.5a | 200-h.p. Hispano-Suiza 200-h.p. Wolseley 'Viper' |
| 85 | 22.5.18 | Fighting (May 1918– Nov. 1918) | S.E.5a | 200-h.p. Wolseley 'Viper' |
| 87 | 26.4.18 | Fighting (Apr. 1918– Nov. 1918) | Sopwith 'Dolphin' | 200-h.p. Hispano-Suiza |
| 88 | 20.4.18 | Fighting Reconnaissance (Apr. 1918– Nov. 1918) | Bristol Fighter | 190–250-h.p. Rolls-Royce 'Falcon' |
| 92 | 1.7.18 | Fighting (July 1918– Nov. 1918) | S.E.5a | 200-h.p. Wolseley 'Viper' |
| 94 | 31.10.18 | Fighting (Oct. 1918– Nov. 1918) | S.E.5a | 200-h.p. Wolseley 'Viper' |

## LIST OF SQUADRONS

| Squadron No. | Date of arrival in France. | Duties. | Aeroplanes. Type. | Engine. |
|---|---|---|---|---|
| 97 | 4.8.18 (Independent Air Force) | Night Bombing (Aug. 1918–Nov. 1918) | Handley Page | 2–250-h.p. Rolls-Royce 'Eagle' |
| 98 | 3.4.18 | Bombing (Apr. 1918–Nov. 1918) | D.H.9 | 200-h.p. B.H.P. |
| 99 | 25.4.18 | Bombing (Apr. 1918–Nov. 1918) | D.H.9<br>D.H.9a | 200-h.p. B.H.P.<br>400-h.p. Liberty |
| 100 | 28.3.17 To 41st Wing 11.10.17 | Night Bombing (Mar. 1917–Nov. 1918) | F.E.2b<br><br>Handley Page | { 120-h.p. Beardmore<br>{ 160-h.p. Beardmore<br>2–250-h.p. Rolls-Royce 'Eagle' |
| 101 | 25.7.17 | Night Bombing (July 1917–Nov. 1918) | F.E.2b | 160-h.p. Beardmore |
| 102 | 24.9.17 | Night Bombing (Sept. 1917–Nov. 1918) | F.E.2b | 160-h.p. Beardmore |
| 103 | 12.5.18 | Bombing (May 1918–Nov. 1918) | D.H.9 | { 200-h.p. B.H.P.<br>{ 230-h.p. Siddeley 'Puma' |
| 104 | 19.5.18 | Bombing (May 1918–Nov. 1918) | D.H.9 | 200-h.p. B.H.P. |
| 107 | 9.6.18 | Bombing (June 1918–Nov. 1918) | D.H.9 | 200-h.p. B.H.P. |
| 108 | 22.7.18 | Bombing (July 1918–Nov. 1918) | D.H.9 | 200-h.p. B.H.P. |
| 110 | 31.8.18 (Independent Air Force) | Bombing (Aug. 1918–Nov. 1918) | D.H.9a | 400-h.p. Liberty |
| 115 | 31.8.18 (Independent Air Force) | Night Bombing (Aug. 1918–Nov. 1918) | Handley Page | 2–250-h.p. Rolls-Royce 'Eagle' |
| 148 | 26.4.18 | Night Bombing (Apr. 1918–Nov. 1918) | F.E.2b | 160-h.p. Beardmore |

# ON WESTERN FRONT 1914–18

| Squadron No. | Date of arrival in France. | Duties. | Aeroplanes. | |
|---|---|---|---|---|
| | | | Type. | Engine. |
| 149 | 2.6.18 | Night Bombing (June 1918–Nov. 1918) | F.E.2b | 160-h.p. Beardmore |
| 151 | 16.6.18 | Night Fighting (June 1918–Nov. 1918) | Sopwith 'Camel' | 110-h.p. Le Rhône |
| 152 | 18.10.18 | Night Fighting (Oct. 1918–Nov. 1918) | Sopwith 'Camel' | 110-h.p. Le Rhône |
| H.Q. Communication Squadron | Apr. 1918 | Communication (Apr. 1918–Nov. 1918) | A.W. R.E.8 | 160-h.p. Beardmore 150-h.p. R.A.F. 4a |
| Special Duty Flight (became 'I' Flight July 1918) | October 1917 | Intelligence (Oct. 1917–Nov. 1918) | Sopwith Scout B.E.12 F.E.2b | 80-h.p. Le Rhône 150-h.p. R.A.F. 4a 160-h.p. Beardmore |
| 'L' Flight | July 1918 | Long Range Artillery | Bristol Fighter | 190-h.p. Rolls-Royce 'Falcon' 200-h.p. Sunbeam 'Arab' |
| 'N' Flight | September 1918 | Long Range Artillery | Bristol Fighter | 190-h.p. Rolls-Royce 'Falcon' 200-h.p. Sunbeam 'Arab' |
| 'P' Flight | October 1918 | Long Range Artillery | Bristol Fighter | 200-h.p. Sunbeam 'Arab' |
| 'M' Flight | 6.10.18 | Long Range Artillery | Bristol Fighter | 190-h.p. Rolls-Royce 'Falcon' |
| 'O' Flight | 1.11.18 | Long Range Artillery | Bristol Fighter | 200-h.p. Sunbeam 'Arab' |
| 17 (American) | 20.6.18 Ceased to be with R.A.F. from 1.11.18 | Fighting (June 1918–Oct. 1918) | Sopwith 'Camel' | 110-h.p. Le Rhône |
| 148 (American) | 1.7.18 Ceased to be with R.A.F. from 1.11.18 | Fighting (July 1918–Oct. 1918) | Sopwith 'Camel' | 130-h.p. Clerget |

# APPENDIX XXIX

## LIST OF NAVAL SQUADRONS WHICH SERVED WITH THE R.F.C. AND R.A.F ON THE WESTERN FRONT 1914-18

(*Note.* On the formation of the R.A.F. on the 1st April 1918, Naval Squadrons were renumbered from No. 201 upwards; for example—No. 1 Squadron R.N.A.S. became No. 201 Squadron R.A.F.)

| Squadron No. | Dates of attachment to R.F.C. | Duties. | Aeroplanes. Type. | Aeroplanes. Engine. |
|---|---|---|---|---|
| 1 (Naval) later 201 | 15.2.17 to 2.11.17 | Fighting | Sopwith Triplane | 130-h.p. Clerget |
|  | 27.3.18 to 11.11.18 | Fighting | Sopwith 'Camel' | 150-h.p. B.R.1 |
| 3 (Naval) later 203 | 1.2.17 to 15.6.17 | Fighting | Sopwith Scout ('Pup') | 80-h.p. Le Rhône |
|  | 3.3.18 to 11.11.18 | Fighting | Sopwith 'Camel' | 150-h.p. B.R.1 |
| 204 | 3.3.18 to 19.4.18 24.10.18 to 11.11.18 | Fighting | Sopwith 'Camel' | 150-h.p. B.R.1 |
| 5 (Naval) later 205 | 3.3.18 to 11.11.18 | Bombing | D.H.4 D.H.9a | 250-h.p. Rolls-Royce 'Eagle' 400-h.p. Liberty |
| 6 (Naval) disbanded 27.8.17 Reformed 1.1.18. Redesignated 206 1.4.18. | 15.3.17 to 26.8.17 | Fighting | Nieuport Scout Sopwith 'Camel' | 130-h.p. Clerget 130-h.p. Clerget |
|  | 31.3.18 to 11.11.18 | Bombing | D.H.9 | 230-h.p. Siddeley 'Puma' |
| 207 | 3.3.18 to 22.4.18 7.6.18 to 11.11.18 | Night Bombing | Handley Page | 2–250-h.p. Rolls-Royce 'Eagle' |

## NAVAL SQUADRONS ON WESTERN FRONT

| Squadron No. | Dates of attachment to R.F.C. | Duties. | Aeroplanes. Type. | Aeroplanes. Engine. |
|---|---|---|---|---|
| 8 (Naval) later 208 | 26.10.16 to 1.2.17 | Fighting | Sopwith Two-seater ('1½-strutter') | 110-h.p. Clerget / 130-h.p. Clerget |
| | | | Sopwith Scout ('Pup') | 80-h.p. Le Rhône |
| | | | Nieuport Scout | 80-h.p. Le Rhône |
| | 27.3.17 to 1.3.18 | Fighting | Sopwith Triplane | 130-h.p. Clerget |
| | | | Sopwith 'Camel' | 130-h.p. Clerget |
| | 2.4.18 to 11.11.18 | Fighting | Sopwith 'Camel' | 150-h.p. B.R.1 / 130-h.p. Clerget |
| | | | Sopwith 'Snipe' | 230-h.p. B.R.2 |
| 9 (Naval) later 209 | 15.6.17 to 28.9.17 | Fighting | Sopwith Triplane | 130-h.p. Clerget |
| | | | Sopwith Scout ('Pup') | 80-h.p. Le Rhône |
| | | | Sopwith 'Camel' | 130-h.p. Clerget |
| | 27.3.18 to 11.11.18 | Fighting | Sopwith 'Camel' | 150-h.p. B.R.1 |
| 10 (Naval) later 210 | 15.5.17 to 20.11.17 | Fighting | Sopwith Triplane | 130-h.p. Clerget |
| | | | Sopwith 'Camel' | 150-h.p. B.R.I. |
| | 27.3.18 to 8.7.18 | Fighting | Sopwith 'Camel' | 150-h.p. B.R.1 |
| | 23.10.18 to 11.11.18 | Fighting | Sopwith 'Camel' | 150-h.p. B.R.1 |
| 211 | 1.4.18 to 11.11.18 | Bombing | D.H.9 | 230-h.p. B.H.P. |
| 214 | 1.4.18 to 11.11.18 | Night Bombing | Handley Page | 2–250-h.p. Rolls-Royce 'Eagle' |
| 215 | 4.7.18 to 11.11.18 | Night Bombing (Independent Air Force) | Handley Page | 2–250-h.p. Rolls-Royce 'Eagle' |
| 'A' Naval Squadron became No. 16 Naval Squadron in Jan. 1918. | 17.10.17 to 11.11.18 | Night Bombing (Independent Air Force) | Handley Page | 2–250-h.p. Rolls-Royce 'Eagle' |

## 144 NAVAL SQUADRONS ON WESTERN FRONT

| Squadron No. | Dates of attachment to R.F.C. | Duties. | Aeroplanes. | |
|---|---|---|---|---|
| | | | Type. | Engine. |
| Later No. 216 Squadron 218 | 17.10.18 to 11.11.18 | Bombing | D.H.9 | 230-h.p. Siddeley 'Puma' |

# APPENDIX XXX

## LOCATION OF ROYAL AIR FORCE UNITS, WESTERN FRONT, 11th NOVEMBER 1918

| | | |
|---|---|---|
| No. 1 Aircraft Depot, H.Q. | Guines | Lieut.-Col. S. A. Hebden |
| No. 1 A.D. (Stores Section) | Guines | Major W. W. Hall |
| No. 1 A.D. (D) | Desvres | Lieut. G. W. Charley |
| No. 1 A.D., M.T. Section | St. Omer | Capt. Phillips |
| Adv. 1 A.D. | Roubaix | |
| No. 1 Aeroplane Supply Depot | Marquise | Lieut.-Col. B. F. Moore |
| No. 1 Issue Section | Rely | Capt. E. Digby Johnson |
| No. 1 Advanced Issue Section | Tourmignies | |
| No. 1 A.S.D. Pool Pilots | Setques | Major J. P. C. Cooper, M.C. |
| No. 1 Aeroplane Repair Park | Marquise | Major J. Kemper, M.B.E. |
| No. 2 Aircraft Depot, H.Q. | Vron | Lieut.-Col. W. H. Lang |
| No. 2 A.D., Stores Section | Cambrai | Major G. Stevens |
| No. 2. A.D., M.T. Section | Cambrai | Major E. F. B. Curtiss |
| Adv. 2 A.D. | Cambrai | |
| No. 2 Aeroplane Supply Depot | Berck-sur-Mer | |
| No. 2 Issue Section | Fienvillers | Capt. A. Denison, M.C. |
| No. 2 A.S.D. Pool Pilots Range | Berck-sur-Mer | Major J. B. Quested, M.C. |
| No. 2 A.S.D. Adv. Pool Pilots | Fienvillers | |
| No. 2 Aeroplane Repair Park. | Bahot | Major J. R. Grant |
| Reception Park | Marquise | Major E. Ainslie, M.B.E. |
| Reinforcement Park | Guines | Major A. N. Stuart |
| Engine Repair Shops | Pont de l'Arche | Col. G. Hynes, D.S.O. |
| Base M.T. Depot | Motteville | Lieut.-Col. A. S. Morris |
| Calais Landing-ground | | |
| Paris office, Hotel La Perouse, Rue la Perouse and Rue Pauquet. | Paris | Capt. A. Wilson |
| Port Depot. | Havre | Lieut. J. S. Done |
| Port Depot. | Boulogne | Lieut. J. M. Patten |
| Port Depot. | Rouen | Lieut.-Col. H. R. Vagg |
| H.Q., Aerodrome Service Unit | Mont St. Eloy | Major H. Ford |

# ROYAL AIR FORCE UNITS, NOV. 1918

G.H.Q., R.A.F.—ST. ANDRE-AUX-BOIS.   ADVANCED G.H.Q., R.A.F. FOSSEUX

'OPERATIONS', ADVANCED G.H.Q., R.A.F.—ESCADŒUVRES

H.Q. Communication Squadron (less 1 Flight) . . . Avesnes-les-Sec
1 Flight, H.Q. Communication Squadron . . . . Berck

### 1st Brigade—Villers Campeaux

G.O.C.:                                     S.O.2.:
Brig.-Gen. D. Le G. Pitcher, C.M.G.   Major M. H. R. K. Hugessen, M.C.

1st (Corps) Wing. Lieut.-Col. E. L. Gossage, M.C.
Valenciennes

| Squadron. | Type. | Corps. | Location. |
|---|---|---|---|
| 5 | R.E.8 | Can. | Aulnoy |
| 16 | R.E.8 | VIII | Auchy |
| 52 | R.E.8 | XXII | Avesnes-les-Sec |
| 'L' Flt. | B.F. | .. | Auberchicourt |

10th (Army) Wing. Lieut.-Col. C. T. McLean, M.C.
Ferme du Muid

| Squadron. | Type. | Location. |
|---|---|---|
| 98 | D.H.9 | Abscon |
| 22 | Bristol Fighter | Aniche |
| 40 | S.E.5a | Aniche |
| 148 | F.E.2b | Erre |
| 203 | 'Camel' (B.R.) | Bruille |
| 'I' Flt. | F.E.2b | Erre |

91st Wing. Lieut.-Col. S. Smith, D.S.O. Rieulay

| Squadron. | Type. | Location. |
|---|---|---|
| 19 | 'Dolphin' | Abscon |
| 64 | S.E.5a | Aniche |
| 209 | 'Camel' (B.R.) | Bruille |

1st Balloon Wing. Lieut.-Col. P. K. Wise, D.S.O.

| Company. | Section. | Corps. |
|---|---|---|
| 1 | .. | XXII |
| .. | 8 | XXII |
| .. | 20 | XXII |
| .. | 28 | XXII |
| 2 | .. | VIII |
| .. | 10 | VIII |
| .. | 24 | VIII |
| .. | 42 | VIII |
| 4 | .. | .. |
| .. | 24 | Can. Corps |
| .. | 46 | Can. Corps |
| 10 | .. | .. |
| .. | 5 | Can. Corps |
| .. | 37 | Can. Corps |

1st Aircraft Park—Somain.
1st Air Ammunition Column—Ferme du Muid.
1st Reserve Lorry Park—Ferme du Muid.
11th Reserve Lorry Park—Ferme du Muid.

## ROYAL AIR FORCE UNITS

### 2ND BRIGADE—ROUBAIX

G.O.C.: Brig.-Gen. T. I. Webb-Bowen, C.M.G.　　S.O.2.: Major C. Beatson

**2nd (Corps) Wing.** Lieut.-Col. A. S. Barratt, M.C. Menin

| Squadron. | Type. | Corps. | Location. |
|---|---|---|---|
| 4 | R.E.8 | XV | Linselles, W. |
| 7 | R.E.8 | II | Menin, E. |
| 10 | A.W.B. | XIX | Staceghem |
| 53 | R.E.8 | X | Sweveghem |
| 82 | A.W.B. | II | Coucou |
| 'M' Flt. | Bristol (Rolls-Arab) | | Menin, W. |

**11th (Army) Wing.** Lieut.-Col. H. A. Van Ryneveld, M.C. Sterhoek

| Squadron. | Type. | Location. |
|---|---|---|
| 29 | S.E.5 | Marcke |
| 41 | S.E.5 | Halluin, E. |
| 48 | Bristol Fighter | Reckem, N. |
| 70 | 'Camel' (Cl.) | Halluin, W. |
| 74 | S.E.5 | Cuerne |
| 79 | 'Dolphin' | Reckem, S. |
| 149 | F.E.2b | St. Marguerite |
| 206 | D.H.9 | Linselles, E. |
| 24 | S.E.5 | Bisseghem |

**65th Wing.** Lieut.-Col. J. A. Cunningham, D.F.C. Heule

| Squadron. | Type. | Location. |
|---|---|---|
| 38 | F.E.2b | Harlebeke |
| 65 | 'Camel' (Cl.) | Bisseghem, S. |
| 108 | D.H.9 | Bisseghem, N. |
| 204 | 'Camel' | Courtrai |
| 39 | Bristol Fighter | Bavichove |

**2nd Balloon Wing.** Lieut.-Col. H. M. Meyler, M.C. Menin

| Company. | Section. | Corps. |
|---|---|---|
| 5 | .. | X |
| .. | 2 | X |
| .. | 25 | X |
| 6 | .. | XV |
| .. | 9 | XV |
| .. | 16 | XV |
| .. | 32 | XV |
| 7 | .. | II |
| .. | 15 | II |
| .. | 38 | II |
| 8 | .. | XIX |
| .. | 13 | XIX |
| .. | 39 | XIX |
| 17 | .. | II |
| .. | 18 | II |
| .. | 36 | II |
| .. | 23 | (Meteor) |

2nd Aircraft Park—Bisseghem.
5th Aircraft Park—Wevelghem.

# WESTERN FRONT, 11TH NOVEMBER 1918

*2nd Balloon Wing* (cont.):
- 2nd Reserve Lorry Park—Menin.
- 7th Reserve Lorry Park—Wevelghem.
- 2nd Air Ammunition Column—Menin.
- 7th Air Ammunition Column—Wevelghem.
- No. 8 Salvage Section—Wevelghem.

### 3RD BRIGADE—RUMILLY

*G.O.C.:*  
Brig.-Gen. C. A. H. Longcroft, D.S.O.

*S.O.2:*  
Major W. S. Evans.

*12th (Corps) Wing.* Lieut.-Col. A. B. Burdett, D.S.O. Ligny-en-Cambresis

| Squadron. | Type. | Corps. | Location. |
|---|---|---|---|
| 12 | R.E.8 | VI | Estourmel |
| 13 | R.E.8 | XVII | Carnier |
| 15 | R.E.8 | V | Selvigny |
| 59 | R.E.8 | IV | Caudry |
| 'N' Flt. | B. Ftr. | .. | Estourmel |

*13th (Army) Wing.* Lieut.-Col. A. J. L. Scott, M.C.
Caudry

| Squadron. | Type. | Location. |
|---|---|---|
| 56 | S.E.5 | La Targette |
| 60 | S.E.5 | Quievy |
| 87 | 'Dolphin' | Boussières |
| 201 | 'Camel' (B.R.I.) | La Targette |
| 210 | 'Camel' | Boussières |

*90th Wing.* Lieut.-Col. G. W. P. Dawes, D.S.O.
Caudry

| Squadron. | Type. | Location. |
|---|---|---|
| 3 | 'Camel' (Le R.) | Inchy |
| 11 | B. Ftr. | Bethencourt |
| 57 | D.H.4 | Bethencourt |
| 102 | F.E.2b | Bevillers |

*3rd Balloon Wing.* Lieut.-Col. G. F. H. Faithfull.
Le Quesnoy

| Company. | Section. | Corps. |
|---|---|---|
| 16 | .. | XVII |
| .. | 12 | XVII |
| .. | 43 | XVII |
| 12 | .. | VI |
| .. | 45 | VI |
| .. | 35 | VI |
| .. | 41 | VI |
| 19 | .. | IV |
| .. | 11 | IV |
| .. | 44 | IV |
| 18 | .. | V |
| .. | 31 | V |
| .. | 1 | V |

3rd Aircraft Park—Bihucourt.
3rd Air Ammunition Column—Ribecourt.

## ROYAL AIR FORCE UNITS

*3rd Balloon Wing* (cont.):

    3rd Reserve Lorry Park—Bihucourt.
    19th Reserve Lorry Park—Bihucourt.
    No. 6 Salvage Section—Caudry.
    No. 9 Salvage Section—Caudry.

### 5TH BRIGADE—SERAIN

*G.O.C.:*                                                     *S.O.2:*
Brig.-Gen. L. E. O. Charlton, C.M.G., D.S.O.

*15th (Corps) Wing.* Lieut.-Col. J. A. Chamier, D.S.O., O.B.E.
                    Elincourt

| Squadron. | Type. | Corps. | Location. |
|---|---|---|---|
| 3rd A.F.C. | R.E.8 | .. | Premont |
| 8 | A.W.B. | Tanks | Malincourt |
| 9 | R.E.8 | IX | Premont |
| 35 | A.W.B. | XIII | Elincourt |
| 73 | 'Camel' (Cl.) | .. | Malincourt |
| 'O' Flt. | B. Ftr. | .. | Premont |

*22nd (Army) Wing.* Lieut.-Col. T. A. E. Cairnes, D.S.O. Reumont

| Squadron. | Type. | Location. |
|---|---|---|
| 46 | 'Camel' (Le R.) | Busigny |
| 80 | 'Camel' (Le R.) | Bertry, W. |
| 84 | S.E.5 | Bertry, W. |
| 208 | 'Camel' (Cl.) | Maretz |

*89th (Army) Wing.* Lieut.-Col. C. E. C. Rabagliati, D.S.O., M.C.
                    Reumont

| Squadron. | Type. | Location. |
|---|---|---|
| 20 | B. Ftr. | Iris Farm |
| 23 | 'Dolphin' | Bertry, E. |
| 92 | S.E.5 | Bertry, E. |
| 101 | F.E.2b | Hancourt |
| 211 | D.H.9 | Iris Farm |
| 218 | D.H.9 | Reumont |

*5th Balloon Wing.* Lieut.-Col. F. M. Roxby. Maroilles

| Company. | Section. | Corps. |
|---|---|---|
| 13 | .. | IX |
| .. | 3 | IX |
| .. | 29 | Reserve |
| 15 | .. | IX |
| .. | 22 | IX |
| .. | 6 | Reserve |
| 14 | .. | XIII |
| .. | 14 | Reserve |
| .. | 19 | XIII |

    4th Aircraft Park—Premont.
    4th Air Ammunition Column—Premont.
    4th Reserve Lorry Park—Premont.
    12th Reserve Lorry Park—Premont.
    No. 7 Salvage Section—Bohain.

# WESTERN FRONT, 11TH NOVEMBER 1918

### 9TH BRIGADE—NOYELLES SUR L'ESCAUT

G.O.C.:  
Brig.-Gen. H. E. Smythe-Osbourne.

S.O.2:  
Major P. C. Hoyland.

*9th Wing.* Lieut.-Col. A. Holt, D.S.O. Marquion

| Squadron. | Type. | Location. |
|---|---|---|
| 25 | D.H.4 | La Brayelles |
| 27 | D.H.9 | Villers-lez-Cagnicourt |
| 32 | S.E.5 | La Brayelles |
| 49 | D.H.9 | Villers-lez-Cagnicourt |
| 62 | B. Ftr. | Villers-lez-Cagnicourt |
| 18 | D.H.4 & D.H.9a | La Brayelles |

*51st Wing.* Lieut.-Col. R. G. Blomfield, D.S.O. Cartigny

| Squadron. | Type. | Location. |
|---|---|---|
| 1 | S.E.5 | Bouvincourt |
| 43 | Sopwith 'Snipe' | Bouvincourt |
| 94 | S.E.5 | Senlis |
| 107 | D.H.9 | Moislains |
| 205 | D.H.4 and D.H. 9a | Moislains |

*54th Wing.* Lieut.-Col. R. G. D. Small. Estrees-en-Chaussée

| Squadron. | Type. | Location. |
|---|---|---|
| 83 | F.E.2b | Estrées-en-Chaussée |
| 151 | 'Camel' (Le R.) | Bancourt |
| 207 | H.P. | Estrées-en-Chaussée |
| 101 | F.E.2b (attd. 5th Bde.) | Hancourt |
| 102 | F.E.2b (attd. 3rd Bde.) | Bevillers |
| 148 | F.E.2b (attd. 1st Bde.) | Erre |
| 149 | F.E.2b (attd. 2nd Bde.) | St. Marguerite |

*82nd Wing.* Lieut.-Col. C. F. Kilner. Carvin

| Squadron. | Type. | Location. |
|---|---|---|
| 58 | H.P. | Provin |
| 152 | 'Camel' (Le R.) | Carvin |
| 214 | H.P. | Chemy |

9th Aircraft Park—Noyelles-sur-l'Escaut.  
5th Air Ammunition Column—Estrées-en-Chaussée.  
9th Air Ammunition Column—Noyelles-sur-l'Escaut.  
6th Reserve Lorry Park  
20th Reserve Lorry Park

### 10TH BRIGADE—LILLE

G.O.C.:  
Brig.-Gen. E. R. Ludlow-Hewitt, D.S.O., M.C.

S.O.2:  
Major R. de Poix.

*80th (Army) Wing.* Lieut.-Col. L. A. Strange, M.C., D.F.C.  
Wattignies

| Squadron. | Type. | Location. |
|---|---|---|
| 2nd A.F.C. | S.E.5 | Pont-à-Marcq |
| 4th A.F.C. | 'Snipe' | Ennetières |
| 88 | B. Ftr. | Bersée |
| 103 | D.H.9 | Ronchin |
| 54 | 'Camel' (Le R.) | Merchin |
| 85 | S.E.5 | Phalempin |

*81st (Corps) Wing.* Lieut.-Col. A. G. R. Garrod, M.C. Sainghin

| Squadron. | Type. | Corps. | Location. |
|---|---|---|---|
| 2 | A.W.B. | I | Genech |
| 21 | R.E.8 | III | Seclin |
| 42 | R.E.8 | XI | Ascq |
| 'P' Flt. | B.F. | .. | Cysoing |
| 6 | R.E.8 | Cav. | Gondecourt |

*8th Balloon Wing.* Lieut.-Col. C. H. Stringer. Cysoing

| Company. | Section. | Corps. |
|---|---|---|
| 3 | .. | I |
| .. | 21 | I |
| .. | 30 | I |
| 11 | .. | XI |
| .. | 40 | XI |
| .. | 47 | XI |
| 20 | .. | III |
| .. | 4 | III |
| .. | 34 | III |
| .. | 48 | III |

10th Aircraft Park—Faubourg-des-Postes, Fonderie Dariel Butin, Rue Marquillies.
10th Air Ammunition Column—Lesquin.
9th Reserve Lorry Park—Faubourg-des-Postes.

## INDEPENDENT FORCE

Head-qu..ters: Autigny La Tour (4m. NE. of Neufchâteau)

*G.O.C.:* Maj.-Gen. Sir Hugh M. Trenchard, K.C.B., D.S.O.

*S.O.1:* Lieut.-Col. E. B. Gordon, D.S.O.

### 8TH BRIGADE—FROVILLE

*G.O.C.:* Brig.-Gen. C. L. N. Newall.   *S.O.2:* Major A. Prout, M.C.

*41st Wing.* Lieut.-Col. J. E. A. Baldwin, D.S.O. Lupcourt

| Squadron. | Type. | Location. |
|---|---|---|
| 55 | D.H.4 | Azelot |
| 99 | D.H.9 | Azelot |
| 104 | D.H.9 | Azelot |

*83rd Wing.* Lieut.-Col. J. H. A. Landon, D.S.O. Xaffrévillers

| Squadron. | Type. | Location. |
|---|---|---|
| 97 | H.P. | Xaffrévillers |
| 100 | H.P. | Xaffrévillers |
| 115 | H.P. | Roville |
| 215 | H.P. | Xaffrévillers |
| 216 | H.P. | Roville |

*88th Wing.* Bettoncourt

| Squadron. | Type. | Location. |
|---|---|---|
| 45 | 'Camel' | Bettoncourt |
| 110 | D.H.9a | Bettoncourt |

3rd Aircraft Depot—Courban.
6th Aircraft Park—Vezelise.

# WESTERN FRONT, 11TH NOVEMBER 1918

*88th Wing* (cont.):
- 12th Aircraft Park—Rambervillers.
- 8th Air Ammunition Column—Vezelise.
- 5th Reserve Lorry Park—Vezelise.
- 10th Reserve Lorry Park—Bulgnéville.
- 3rd Aeroplane Supply Depot—Courban.

### SQUADRONS
*In Numerical Order*

| Number. | Brigade. | Aerodrome. | Commanding Officer. |
|---|---|---|---|
| 1 | 9th Brigade | Bouvincourt | Major W. E. Young |
| 2 | 10th Brigade | Genech | Major P. G. Ross-Hume, M.C. |
| 2nd A.F.C. | 10th Brigade | Pont-à-Marq | Major A. Murray-Jones, M.C., D.F.C. |
| 3 | 3rd Brigade | Inchy | Major R. St. Clair McClintock |
| 3rd A.F.C. | 5th Brigade | Premont, S. | Major W. H. Anderson, D.F.C. |
| 4 | 2nd Brigade | Linselles | Major R. E. Saul |
| 4th A.F.C. | 10th Brigade | Ennetières | Major W. A. McClaughry, M.C., D.F.C. |
| 5 | 1st Brigade | Aulnoy | Major G. Knight, M.C. |
| 6 | 10th Brigade | Gondecourt | Major G. C. Pirie, M.C. |
| 7 | 2nd Brigade | Menin, E. | Major B. E. Sutton, D.S.O., M.C. |
| 8 | 5th Brigade | Malincourt | Major T. B. Leigh-Mallory |
| 9 | 5th Brigade | Premont | Major J. T. Rodwell |
| 10 | 2nd Brigade | Staceghem | Major K. D. P. Murray, M.C. |
| 11 | 3rd Brigade | Bethencourt | Major R. W. Heath |
| 12 | 3rd Brigade | Estourmel | Major T. Q. Back |
| 13 | 3rd Brigade | Carnieres | Major A. P. D. Hill |
| 15 | 3rd Brigade | Selvigny | Major H. V. Stammers, D.F.C. |
| 16 | 1st Brigade | Auchy | Major A. W. C. V. Parr |
| 18 | 9th Brigade | La Brayelles | Major J. F. Gordon, D.F.C. |
| 19 | 1st Brigade | Abscon | Major H. W. G. Jones, M.C. |
| 20 | 5th Brigade | Iris Farm | Major E. H. Johnston |
| 21 | 10th Brigade | Seclin | Major A. A. Walser, M.C. |
| 22 | 1st Brigade | Aniche | Major J. McKelvie |
| 23 | 5th Brigade | Bertry, E. | Major C. E. Bryant, D.S.O. |
| 24 | 2nd Brigade | Bisseghem | Major V. A. H. Robeson, M.C. |
| 25 | 9th Brigade | La Brayelles | Major C. S. Duffus, M.C. |
| 27 | 9th Brigade | Villers-lez-Cagnicourt | Major G. D. Hill |
| 29 | 2nd Brigade | Marcke | Major C. H. Dixon, M.C. |
| 32 | 9th Brigade | La Brayelles | Major J. C. Russell |
| 35 | 5th Brigade | Elincourt | Major F. D. Stevenson, D.S.O., M.C. |
| 38 | 2nd Brigade | Harlebeke | Major R. D. Oxland |
| 39 | 2nd Brigade | Bavichove | Major W. T. F. Holland |
| 40 | 1st Brigade | Aniche | Major R. J. Compston, D.S.C., D.F.C. |
| 41 | 2nd Brigade | Halluin, E. | Major G. H. Bowman, D.S.O., M.C. |
| 42 | 10th Brigade | Ascq | Major H. J. F. Hunter, M.C. |
| 43 | 9th Brigade | Bouvincourt | Major C. C. Miles, M.C. |

152 ROYAL AIR FORCE UNITS

| Number. | Brigade. | Aerodrome. | Commanding Officer. |
|---|---|---|---|
| 46 | 5th Brigade | Busigny | Major G. Allen |
| 48 | 2nd Brigade | Reckem | Major K. R. Park, M.C. |
| 49 | 9th Brigade | Villers-lez-Cagnicourt | Major B. S. Benning |
| 52 | 1st Brigade | Avesnes les Sec | Major A. M. Morison |
| 53 | 2nd Brigade | Sweveghem | Major G. Henderson |
| 54 | 10th Brigade | Merchin | Major R. S. Maxwell, M.C. |
| 55 | Independent Force | Azelot | Major A. Gray, M.C. |
| 56 | 3rd Brigade | La Targette | Major E. J. L. Gilchrist, M.C. |
| 57 | 3rd Brigade | Bethencourt | Major G. C. Bailey, D.S.O. |
| 58 | 9th Brigade | Provin | Major D. Gilley, D.F.C. |
| 59 | 3rd Brigade | Caudry | Major C. J. Mackay, M.C. |
| 60 | 3rd Brigade | Quievy | Major A. C. Clarke |
| 62 | 9th Brigade | Villers-lez-Cagnicourt | Major F. W. Smith |
| 64 | 1st Brigade | Aniche | Major B. E. Smythies |
| 65 | 2nd Brigade | Bisseghem | Major H. V. Champion de Crespigny, M.C., D.F.C. |
| 70 | 2nd Brigade | Halluin, W. | Major G. W. M. Green, D.S.O., M.C. |
| 73 | 5th Brigade | Hervilly | Major M. Le Blanc-Smith, D.F.C. |
| 74 | 2nd Brigade | Cuerne | Major K. L. Caldwell, M.C., D.F.C. |
| 79 | 2nd Brigade | Reckem | Major A. R. Arnold, D.S.C. |
| 80 | 5th Brigade | Bertry, W. | Major D. V. Bell |
| 82 | 2nd Brigade | Coucou | Major J. B. Solomon, M.C. |
| 83 | 9th Brigade | Estrées-en-Chaussée | Major S. W. Price, M.C. |
| 84 | 5th Brigade | Bertry, W. | Major G. E. M. Pickthorne |
| 85 | 10th Brigade | Phalempin | Major C. M. Crowe, M.C., D.F.C. |
| 87 | 3rd Brigade | Boussières | Major C. J. W. Darwin, D.S.O. |
| 88 | 10th Brigade | Bersée | Major R. N. M. Stuart-Wortley, M.C. |
| 92 | 5th Brigade | Bertry, E. | Major A. Coningham, D.S.O., M.C. |
| 94 | 9th Brigade | Senlis | Major A. J. Capel |
| 97 | Independent Force | Xaffrévillers | Major V. A. Albrecht |
| 98 | 1st Brigade | Abscon | Major P. C. Sherren, M.C. |
| 99 | Independent Force | Azelot | Major L. A. Pattinson, M.C. |
| 100 | Independent Force | Xaffrévillers | Major C. G. Burge |
| 101 | 5th Brigade | Hancourt | Major J. Sowrey, A.F.C. |
| 102 | 3rd Brigade | Bevillers | Major F. C. Baker |
| 103 | 10th Brigade | Ronchin | Major M. H. B. Nethersole |
| 104 | Independent Force | Azelot | Major J. C. Quinnell |
| 107 | 9th Brigade | Moislains | Major H. G. Dean |

# WESTERN FRONT, 11TH NOVEMBER 1918

| Number. | Brigade. | Aerodrome. | Commanding Officer. |
|---|---|---|---|
| 108 | 2nd Brigade | Bisseghem | Major B. F. Vernon-Harcourt |
| 110 | Independent Force | Bettoncourt | Major H. R. Nicholl |
| 115 | Independent Force | Roville | Major W. E. Gardner |
| 148 | 1st Brigade | Erre | Major I. T. Lloyd |
| 149 | 2nd Brigade | St. Marguerite | Major P. B. Greenwood |
| 151 | 9th Brigade | Bancourt | Major C. J. Q. Brand, D.S.O., M.C., D.F.C. |
| 152 | 9th Brigade | Carvin | Major E. Henty |
| 201 | 3rd Brigade | La Targette | Major C. M. Leman |
| 203 | 1st Brigade | Bruille | Major T. F. Hazell, D.S.O., M.C., D.F.C. |
| 204 | 2nd Brigade | Courtrai | Major E. W. Norton, D.S.C. |
| 205 | 9th Brigade | Moislains | Major J. B. Elliot |
| 206 | 2nd Brigade | Linselles, E. | Major C. T. Maclaren |
| 207 | 9th Brigade | Estrées-en-Chaussée | Major G. R. Elliott |
| 208 | 5th Brigade | Maretz | Major C. Draper, D.S.C. |
| 209 | 1st Brigade | Bruille | Major T. F. W. Gerrard, D.S.C. |
| 210 | 3rd Brigade | Boussières | Major B. C. Bell, D.S.O., D.S.C. |
| 211 | 5th Brigade | Iris Farm | Major G. R. M. Reid, M.C. |
| 214 | 9th Brigade | Chemy | Major H. G. Brackley, D.S.O., D.S.C. |
| 215 | Independent Force | Xaffrévillers | Major J. F. Jones |
| 216 | Independent Force | Autreville | Major H. O. Buss, D.S.C. |
| 218 | 5th Brigade | Reumont | Major B. S. Wemp, D.F.C. |
| 'I' Flt. (F.E.2b) | 1st Brigade | Erre | Captain P. W. B. Lawrence |
| 'L' Flt. (B.F.) | 1st Brigade | Auberchicour | Captain B. P. Catchpole, M.C., D.F.C. |
| 'M' Flt. (B.F.) | 2nd Brigade | Menin | Captain A. W. F. Glenny, M.C., D.F.C. |
| 'N' Flt. (B.F.) | 3rd Brigade | Estourmel | Captain C. E. Barrington |
| 'O' Flt. (B.F.) | 5th Brigade | Premont | Captain E. J. Jones, M.C., D.F.C. |
| 'P' Flt. (B.F.) | 10th Brigade | Cysoing | Captain L. H. Jones, D.F.C. |

# APPENDIX XXXI

## BRITISH AIRCRAFT PRODUCED AND LABOUR EMPLOYED AUGUST 1914 TO NOVEMBER 1918

(*Figures for Germany, France, Italy, and America given, where available, for comparison*)

### First-line Strength in Aircraft

August 1914
- Naval . . . . . 50 aeroplanes and seaplanes
  6 airships
- Military . . . . . 63 aeroplanes

November 1918   Royal Air Force   3,300 aeroplanes and seaplanes
   103 airships

(*Note.* In August 1914 France had 120 aeroplanes ready to take the field,[1] and Germany 232 aeroplanes.[2] The French first-line strength at the armistice was 4,511 aeroplanes,[1] and the German 2,390.[3])

### Aircraft Production 1914–18

Total airframes manufactured in Great Britain . . . . 55,093
Total aircraft engines manufactured in Great Britain . . . 41,034
Airframes purchased abroad . . . . . . . 3,051
Aircraft engines purchased abroad . . . . . . 16,897

(*Note.* France built a total of 67,982 airframes and 85,317 engines.[4] Italy is said to have manufactured 20,000 airframes and 38,000 engines.[5] In the 21 months of her participation in the War, America produced for her own use 15,000 aeroplanes and 41,000 engines.[6] German production up to January 1919 was 47,637 airframes and 40,449 engines.[7])

### Average Monthly Output, 1917 and 1918

|  | 1917 | 1918 |
|---|---|---|
| Airframes | 1,229 | 2,668 |
| Engines | 980 | 1,841 |

(*Note.* For France the comparable figures are: *1917*, airframes 1,894, engines 1,734; *1918*, airframes 2,852, engines 3,359.[8])

---

[1] *La D.C.A. (Défense contre Aéronefs)*, by Chef d'Escadron Jean Lucas, p. 14.
[2] *Deutschlands Krieg in der Luft*, by General von Hoeppner, p. 7.
[3] Figures supplied by the *Reichsarchiv*, Potsdam. The strength is, however, nominal, not actual. The Germans usually estimate their actual at about two-thirds of their nominal strength.
[4] *La D.C.A.*, p. 13.   [5] *La Revue des Forces Aériennes*, October 1930.
[6] *Aero Digest*, January 1934.
[7] *Deutschlands Luftstreitkräfte im Weltkriege*, by Major G. P. Neumann, p. 585.
[8] *La D.C.A.*, p. 14.

# AIRCRAFT AND LABOUR

*Total Strength in Aircraft, 31st October 1918*

| | |
|---|---:|
| Airframes | 22,171 |
| Engines | 37,702 |

(*Note.* The German strength in aircraft was, very approximately, 20,000 at the armistice. A French official figure for August 1918 gave the French strength in serviceable aircraft as a total of 15,342.)

*Labour Employed, October 1918*

| | Men. | Women. | Boys. |
|---|---:|---:|---:|
| On the production of airframes, engines, accessories, and spares | 187,526 | 126,544 | 33,042 |

TOTAL: 347,112

Staff of the Department of Aircraft Production, Ministry of Munitions 17,250

(*Note.* In 1918 the French had 186,000 employees engaged in aircraft production[1] and the Germans about 100,000.[2])

# APPENDIX XXXII

## PRICE-LIST OF VARIOUS BRITISH AIRFRAMES AND ENGINES

(*Extracted from a Ministry of Munitions 'Priced Vocabulary of Aircraft Supplies, 1918–19'*)

### AEROPLANES

(*Note.* The prices do not include the engine, instruments, or guns.)

| Type. | £ | s. | d. |
|---|---:|---:|---:|
| Armstrong Whitworth (F.K.3 90-h.p. R.A.F.) | 1,127 | 10 | 0 |
| Armstrong Whitworth (F.K.8 160-h.p. Beardmore) | 1,365 | 17 | 0 |
| Avro 504k | 898 | 19 | 0 |
| B.E.2c | 1,072 | 10 | 0 |
| B.E.2e | 1,072 | 10 | 0 |
| B.E.12 | 990 | 0 | 0 |
| Bristol Fighter F.2b | 1,350 | 19 | 0 |
| Bristol Monoplane | 770 | 0 | 0 |
| Bristol Scout | 1,072 | 10 | 0 |
| De Havilland 1 | 1,100 | 0 | 0 |
| De Havilland 4 R.A.F. 3a | 1,424 | 10 | 0 |
| De Havilland 5 | 874 | 0 | 0 |
| De Havilland 6 (Curtiss Ox 5) | 885 | 10 | 0 |
| De Havilland 6 R.A.F. | 841 | 10 | 0 |
| De Havilland 9 | 1,473 | 5 | 0 |
| De Havilland 9a | 1,599 | 12 | 0 |
| De Havilland 10 (Liberty Engine) | 3,483 | 7 | 0 |

[1] *La D.C.A.*, p. 14.
[2] According to a statement by Comdt. de Castelnau in his introduction to *L'Allemagne et la Guerre de l'air*, the French translation of General von Hoeppner's *Deutschlands Krieg in der Luft*.

# PRICE-LIST OF VARIOUS

| Type. | Price |
|---|---|
|  | £ s. d. |
| F.E.2b | 1,521 13 4 |
| F.E.2d | 1,540 0 0 |
| Handley Page O.400 | 6,000 0 0 |
| Handley Page V.1,500 | 12,500 0 0 |
| Henri Farman | 1,159 19 0 |
| Martinsyde F.3 (300-h.p. Hispano) | 1,210 0 0 |
| Martinsyde F.3 (Rolls-Royce 'Falcon' III) | 1,155 0 0 |
| Martinsyde F.4 | 1,142 2 0 |
| Maurice Farman (Shorthorn) | 1,005 8 0 |
| R.E.7 | 1,886 10 0 |
| R.E.8 (R.A.F. 4a) | 1,232 0 0 |
| S.E.5a | 1,063 10 0 |
| Sopwith 'Camel' F.1 (110-h.p. Le Rhône, 130-h.p. Clerget, or B.R.1) | 874 10 0 |
| Sopwith Scout (80-h.p. Le Rhône) | 710 18 0 |
| Sopwith 'Dolphin' | 1,010 13 0 |
| Sopwith 'Snipe' (B.R.2) | 945 17 0 |
| Sopwith 'Salamander' | 1,138 0 0 |
| Sopwith 1½ Strutter | 842 6 0 |
| S.P.A.D. | 1,045 0 0 |
| Vickers Bomber F.B.27a Mk.11 | 5,145 0 0 |

### SEAPLANES AND SHIPS' AEROPLANES

(*Note.* The prices do not include the engine, instruments, or guns.)

| Type. | Price. |
|---|---|
|  | £ s. d. |
| Blackburn Twin | 5,648 0 0 |
| Fairey 'Campania' | 3,245 0 0 |
| Hamble 'Baby' (110-h.p. and 130-h.p. Clerget) | 1,175 0 0 |
| Short 184 (225-h.p., 240-h.p., 260-h.p. Sunbeam and 240-h.p. Renault) | 3,107 10 0 |
| Short 320 | 3,589 7 0 |
| Sopwith 'Baby' | 1,072 10 0 |
| Sopwith 2F1 (Ship's aeroplane) | 825 0 0 |
| Sopwith 'Pup' (Ship's aeroplane) | 770 0 0 |
| Sopwith Torpedo Aeroplane | 1,613 10 0 |
| Wight | 2,970 0 0 |

### Flying-boats.

| | |
|---|---|
| A.D. Type (including hull) | 2,853 8 0 |
| A.D. Type (excluding hull) | 1,925 0 0 |
| F.2a Type (including hull and trolley) | 6,738 0 0 |
| F.2a Type (excluding hull but including trolley) | 5,198 0 0 |
| N.T.2b Type | 1,477 0 0 |
| N.T.4 Type 'Small America' | 3,610 0 0 |

### Balloons.

| | |
|---|---|
| Cacquot. Army 'R' type | 400 5 6 |
| Cacquot. Naval 'M' type | 458 6 4 |

# BRITISH AIRFRAMES AND ENGINES

ENGINES

| Type. | Normal H.P. | Number of Cylinders. | Price. | | |
|---|---|---|---|---|---|
| | | | £ | s. | d. |
| Adriatic (B.H.P.) | 230 | 6 | 1,089 | 0 | 0 |
| Atlantic (Galloway) | 500 | 12 | 2,073 | 10 | 0 |
| Anzani | 100 | 10 | 550 | 0 | 0 |
| Arab (Sunbeam) | 200 | 8 | 1,017 | 10 | 0 |
| Adder (Hispano Wolseley W.4b) | 200 | 8 | 946 | 0 | 0 |
| B.R.1 | 150 | 9 | 643 | 10 | 0 |
| B.R.2 | 200 | 9 | 880 | 0 | 0 |
| Beardmore | 120 | 6 | 825 | 0 | 0 |
| Beardmore | 160 | 6 | 1,045 | 0 | 0 |
| Clerget (7.Z) | 80 | 7 | 484 | 0 | 0 |
| Clerget (9b) and (9bf) | 130 | 9 | 907 | 10 | 0 |
| Clerget (11eb) | 200 | 11 | 951 | 0 | 0 |
| Curtiss (Ox. 5) | 90 | 8 | 693 | 10 | 0 |
| Dragonfly (ABC) | 320 | 9 | 1,072 | 0 | 0 |
| Eagle II, III, and IV (Rolls-Royce) | 250 | 12 | 1,430 | 0 | 0 |
| Eagle V (Rolls-Royce) | 275 | 12 | 1,721 | 10 | 0 |
| Eagle VI and VII (Rolls-Royce) | 350 | 12 | 1,919 | 10 | 0 |
| Eagle VIII (Rolls-Royce) | 360 | 12 | 1,622 | 10 | 0 |
| Falcon I, II, and III (Rolls-Royce) | 190 | 12 | 1,210 | 0 | 0 |
| Falcon IV (Rolls-Royce) | 225 | 12 | 1,617 | 0 | 0 |
| F.I.A.T. (A.12 Bis.) | 300 | 6 | 1,617 | 0 | 0 |
| Gnome | 80 | 7 | 430 | 0 | 0 |
| Hawk (Rolls-Royce) | 75 | 6 | 896 | 10 | 0 |
| Hispano Suiza (French) | 200 | 8 | 1,004 | 0 | 0 |
| Lion (Napier) | 400 | 12 | 1,897 | 10 | 0 |
| Liberty | 400 | 12 | 1,215 | 0 | 0 |
| Le Rhône | 110 | 9 | 771 | 10 | 0 |
| Le Rhône | 80 | 9 | 620 | 0 | 0 |
| Monosoupape | 100 | 9 | 696 | 0 | 0 |
| Maori (Sunbeam) | 250 | 12 | 1,391 | 10 | 0 |
| Pacific | 600 | 12 | 2,145 | 0 | 0 |
| Puma (B.H.P.) | 230 | 6 | 1,089 | 0 | 0 |
| R.A.F. 1a | 100 | 8 | 522 | 10 | 0 |
| R.A.F. 3a | 200 | 12 | 1,210 | 0 | 0 |
| R.A.F. 4a | 150 | 12 | 836 | 0 | 0 |
| Renault | 80 | 8 | 522 | 10 | 0 |
| Salmson | 200 | 9 | 968 | 0 | 0 |
| Viper (Hispano Wolseley) W.4a | 200 | 8 | 814 | 0 | 0 |

## APPENDIX XXXIII

### FIRMS AND LABOUR EMPLOYED ON BRITISH AIRCRAFT PRODUCTION (EXCLUDING AIRSHIPS) COMPARATIVE DETAILED STATEMENT FOR THE YEARS 1916, 1917, AND 1918

| | August 1916. | | | | | | November 1917. | | | | | | October 1918. | | | | | |
|---|---|---|---|---|---|---|---|---|---|---|---|---|---|---|---|---|---|---|
| | No. of firms. | % of dilution. | Total. | Men. | Women. | Boys. | No. of firms. | % of dilution. | Total. | Men. | Women. | Boys. | No. of firms. | % of dilution. | Total. | Men. | Women. | Boys. |
| Aeroplanes | 27 | 24·0 | 18,809 | 14,288 | 3,032 | 1,489 | 59 | 36·7 | 60,580 | 38,301 | 17,768 | 4,511 | 64 | 38·1 | 94,700 | 58,568 | 29,138 | 6,994 |
| Seaplanes | 21 | 26·1 | 3,576 | 2,640 | 585 | 351 | 38 | 34·4 | 10,905 | 7,147 | 2,800 | 958 | 58 | 35·6 | 17,685 | 11,389 | 4,976 | 1,320 |
| Propellers | 21 | 19·9 | 887 | 710 | 84 | 93 | 34 | 24·8 | 3,078 | 2,312 | 512 | 254 | 46 | 28·6 | 5,338 | 3,814 | 1,118 | 406 |
| Aero-parts | 88 | 43·5 | 5,855 | 3,290 | 1,799 | 766 | 176 | 50·2 | 21,893 | 10,887 | 8,614 | 2,392 | 331 | 49·0 | 51,011 | 26,013 | 19,538 | 5,460 |
| Engines | 114 | 30·7 | 19,692 | 13,633 | 3,711 | 2,348 | 201 | 38·2 | 57,549 | 35,537 | 15,590 | 6,422 | 323 | 42·7 | 99,362 | 56,850 | 31,852 | 10,660 |
| Total. | 271 | 29·2 | 48,819 | 34,561 | 9,211 | 5,047 | 508 | 38·8 | 154,005 | 94,184 | 45,284 | 14,537 | 822 | 41·9 | 268,096 | 156,634 | 86,622 | 24,840 |
| Materials | 220 | 42·1 | 11,254 | 6,413 | 3,404 | 1,437 | 263 | 50·3 | 19,964 | 9,918 | 7,450 | 2,596 | 707 | 60·9 | 79,016 | 30,892 | 39,922 | 8,202 |
| Grand Total | 491 | 31·7 | 60,073 | 40,974 | 12,615 | 6,484 | 771 | 40·1 | 173,969 | 104,102 | 52,734 | 17,133 | 1,529 | 46·0 | 347,112 | 187,526 | 126,544 | 33,042 |

*Comparative position as at October 1918 of 508 Firms represented in return of November 1917.*

| | No. of firms. | % of dilution. | Total. | Men. | Women. | Boys. |
|---|---|---|---|---|---|---|
| Seaplanes | 38 | 35·6 | 15,313 | 9,856 | 4,336 | 1,121 |
| Propellers | 34 | 28·6 | 4,182 | 2,985 | 889 | 308 |
| Aeroplanes | 59 | 37·5 | 84,805 | 52,926 | 25,289 | 6,590 |
| Aero-parts | 176 | 50·2 | 35,175 | 17,506 | 13,889 | 3,770 |
| Engines | 201 | 42·4 | 87,808 | 50,560 | 27,897 | 9,351 |
| Total | 508 | 41·1 | 227,283 | 133,833 | 72,300 | 21,140 |

# APPENDIX XXXIV

## BRITISH NAVAL AIRSHIPS BUILT 1914–18

| Type | Constructed | | | | | | Transfd. to Allied Governments. | | | | | | Deleted | | | | | | Remaining in commission 1st Nov. 1918. |
|---|---|---|---|---|---|---|---|---|---|---|---|---|---|---|---|---|---|---|---|
| | Before outbreak of war. | 1914. | 1915. | 1916. | 1917. | To 31 Oct. 1918. | Before war. | 1914. | 1915. | 1916. | 1917. | To 31 Oct. 1918. | Before war. | 1914. | 1915. | 1916. | 1917. | To 31 Oct. 1918. | |
| Various | 6 (a) | .. | .. | .. | .. | .. | .. | .. | .. | .. | .. | .. | .. | .. | 2 | 6 | .. | 2 | Nil |
| Rigid | 1 | 1 (h) | 1 (k) | .. | 4 | 3 | .. | .. | .. | .. | .. | .. | 1 | .. | .. | .. | .. | .. | 5 |
| Parseval | 1 | .. | 1 | .. | 2 | .. | .. | .. | .. | .. | .. | .. | .. | .. | .. | .. | 1 | 2 | 1 |
| N. Sea | .. | .. | .. | .. | 5 | 7 | .. | .. | .. | .. | .. | .. | .. | .. | .. | .. | .. | 6 | 6 |
| C. Star | .. | .. | .. | .. | .. | 10 | .. | .. | .. | .. | .. | .. | .. | .. | .. | .. | .. | .. | 10 |
| Coastal | .. | .. | .. | 35 (b) | .. | 13 (c)(d) | .. | .. | .. | 5 | .. | .. | .. | .. | .. | 3 | 9 | 14 | 4 |
| S.S.T. | .. | .. | .. | .. | .. | 46 (e) | .. | .. | .. | .. | 3 | 1 | .. | .. | .. | .. | .. | 1 | 12 (d) |
| S.S.Z. | .. | .. | .. | 1 | 24 | .. | .. | .. | 2 | 6 | 6 | .. | .. | .. | .. | .. | 3 | 11 | 53 (e) |
| S.S.P. | .. | .. | .. | .. | 6 | .. | .. | .. | .. | .. | .. | .. | .. | .. | .. | .. | 3 | .. | 3 |
| S.S. | .. | .. | 29 | 22 (f) | (g)5 | 3 (g) | .. | .. | .. | .. | .. | .. | .. | .. | 4 | 5 | 9 | 18 | 9 |
| | 8 | 1 | 31 | 58 | 46 | 82 | .. | .. | 2 | 11 | 9 | 1 | 1 | .. | 6 | 14 | 25 | 54 | 103 |
| | 226 | | | | | | 23 | | | | | | 100 | | | | | | 103 |

(a) Includes 'Beta', 'Gamma', 'Delta', and 'Eta'. Transferred from the Army. No. 2 Improved Willows. No. 3 Astra Torres.
(b) Including 3 reconstructed ships.
(c) Excluding S.S.T 6 (Deleted). Never accepted and S.S.E. 3 not accepted.
(d) Including S.S.T. Experimental ship.
(e) Including one S.S.Z. in reserve at Kassandra.
(f) Includes 2 reconstructed ships.
(g) Reconstructed ships.
(h) No. 8 Astra Torres.
(k) No. 10 Astra Torres.

Total constructed . . 213
Reconstructed . . 13
                    226

Transferred to Allied Governments . . 23
Deleted . . 100
              123

Balance in Commission 1st Nov. 1918 . . 103

# APPENDIX XXXV

## STRENGTH OF BRITISH AIR PERSONNEL AUGUST 1914 AND NOVEMBER 1918

*August 1914*

Army { 146 Officers
1,097 Other Ranks
Naval { 130 Officers
700 Other Ranks

2,073 Officers and Men

*November 1918*

27,333 Officers
16,681 Cadets and N.C.O.s under instruction
247,161 N.C.O.s and Men

291,175

# APPENDIX XXXVI

## TOTAL CASUALTIES, ALL CAUSES, TO AIR SERVICE PERSONNEL, BRITISH AND GERMAN, 1914–18

|  | British. | | | German.* | | |
|---|---|---|---|---|---|---|
|  | Officers. | N.C.O.s and men. | Total. | Officers. | N.C.O.s and men. | Total. |
| Killed or died | 4,579 | 1,587 | 6,166 | 2,397 | 3,456 | 5,853 |
| Wounded or injured | 5,369 | 1,876 | 7,245 | 3,129 | 4,173 | 7,302 |
| Missing or interned | 2,839 | 373 | 3,212 | 1,364 | 1,387 | 2,751 |
|  | 12,787 | 3,836 | 16,623 | 6,890 | 9,016 | 15,906 |

\* Figures supplied by the *Reicharchiv*, Potsdam.

# APPENDIX XXXVII

COMPARISON, BY MONTHS, OF BRITISH FLYING CASUALTIES (KILLED AND MISSING) AND HOURS FLOWN: WESTERN FRONT, JULY 1916 TO JULY 1918

*(Statement prepared by the Air Ministry for the War Cabinet)*

| Month. | Casualties (killed and missing). | Hours flown. | Hours flown per casualty (killed and missing). |
|---|---|---|---|
| *1916.* | | | |
| July | 75 | 17,000 | 226 |
| August | 66 | 19,500 | 295 |
| September | 105 | 22,500 | 215 |
| October | 75 | 13,500 | 186 |
| November | 59 | 10,600 | 179 |
| December | 39 | 5,200 | 133 |
| | | | 1,234 |
| | | | Average for 6 months 206 |
| *1917.* | | | |
| January | 41 | 10,500 | 256 |
| February | 65 | 12,000 | 183 |
| March | 143 | 14,500 | 101 |
| April | 316 | 29,500 | 92 |
| May | 187 | 39,500 | 211 |
| June | 165 | 35,500 | 215 |
| | | | 1,058 |
| | | | Average for 6 months 176 |
| July | 168 | 32,750 | 194 |
| August | 183 | 31,750 | 173 |
| September | 214 | 38,500 | 179 |
| October | 198 | 30,500 | 154 |
| November | 153 | 17,500 | 114 |
| December | 81 | 18,000 | 222 |
| | | | 1,036 |
| | | | Average for 6 months 172 |
| *1918.* | | | |
| January | 96 | 22,000 | 229 |
| February | 91 | 20,500 | 225 |
| March | 245 | 40,000 | 163 |
| April | 194 | 31,250 | 161 |
| May | 240 | 63,325 | 263 |
| June | 211 | 61,140 | 289 |
| | | | 1,330 |
| | | | Average for 6 months 222 |
| July to 21st | 200 | 41,272 | 206 |

# APPENDIX
## DELIVERIES OF ANTI-AIRCRAFT GUNS AN[D]
*(Extracted from a Min[...]*

|  | 1916. | | | | | |
|---|---|---|---|---|---|---|
|  | 1st qtr. | 2nd qtr. | 3rd qtr. | 4th qtr. | 1st qtr. | 2nd qtr. |
| *Anti-Aircraft Guns (New)* | | | | | | |
| 13-pdr. 9-cwt. I–IV | 26 | 112 | 69 | 31 | 68 | 5 |
| 3-in. 5-cwt. | 2 | .. | .. | .. | .. | .. |
| 3-in. 20-cwt. I–III. | .. | 20 | 52 | 89 | 15 | 2 |
| 4-in. A.A. | .. | .. | .. | .. | .. | .. |
| Total | 28 | 132 | 121 | 120 | 83 | 8 |
| *Anti-Aircraft Guns (Repaired)* | | | | | | |
| 13-pdr. 9-cwt. | .. | .. | .. | 4 | 10 | 2 |
| 3-in. 5-cwt. | .. | .. | .. | .. | .. | .. |
| 3-in. 20-cwt. | .. | .. | .. | .. | .. | .. |
| Total | .. | .. | .. | 4 | 10 | 2 |
| *Anti-Aircraft Ammunition* | | | | | | |
| 6-pdr. / 12-pdr. / 12-cwt. / (3-in.) shell — H.E. | 2,400 | .. | 22,600 | 4,600 / 4,100 | 12,000 | .. |
| S. | .. | .. | .. | .. | .. | .. |
| Incy. | .. | .. | .. | .. | 500 | .. |
| 13-pdr. (6-cwt. or 3-in. shell) — H.E. | 4,000 | 41,800 | 19,200 | 10,100 | 9,200 | 32,20[0] |
| S. | .. | .. | .. | .. | 300 | .. |
| Incy. | .. | .. | .. | .. | .. | .. |
| 13-pdr. 9-cwt. (13-pdr. or 3-in. shell) — H.E. | 1,000 | 18,200 | 111,800 | 154,400 | 173,300 | 224,40[0] |
| S. | 6,500 | 74,700 | 187,400 | 258,300 | 83,100 | 77,60[0] |
| Incy. | .. | .. | .. | .. | 1,000 | 2,90[0] |
| Star. | .. | .. | .. | .. | .. | .. |
| 3-in. 5-cwt. — H.E. | .. | .. | .. | .. | 1,400 | 3,70[0] |
| Incy. | .. | .. | .. | .. | .. | .. |
| 3-in. 20-cwt. (3-in. or 13-pdr. shell) — H.E. | .. | 13,900 | 65,500 | 92,300 | 27,300 | .. |
| S. | 200 | 6,900 | 33,900 | 25,600 | 30,500 | .. |
| Incy. | .. | .. | .. | .. | 14,100 | 3,30[0] |
| Star. | .. | .. | .. | .. | .. | .. |
| 18-pdr. — H.E. | .. | 10,200 | 28,000 | 6,800 | 8,000 | 20,80[0] |
| S. | .. | .. | .. | .. | .. | .. |
| Incy. A.Z. | .. | .. | .. | .. | 2,700 | .. |
| 4-in. Mk. IV. H.E. | .. | .. | .. | .. | .. | .. |
| 4-in. Mk. V. H.E. | .. | .. | .. | .. | .. | .. |
| Total | 14,100 | 165,700 | 468,400 | 556,200 | 363,400 | 364,90[0] |

[1] Includes some ordinary 18-pdr.

## XXXVIII

### AMMUNITION (EXCLUDING NAVAL), 1916–18

*y of Munitions paper)*

| 17. | | 1918. | | | | | |
|---|---|---|---|---|---|---|---|
| 3rd qtr. | 4th qtr. | 1st qtr. | 2nd qtr. | 3rd qtr. | Oct. and 1–11 Nov. | 12 Nov. to 30 Dec. | Total. |
| 51 | 117 | 46 | 69 | 64 | 6 | 6 | 722 |
| .. | .. | .. | .. | .. | .. | .. | 2 |
| 23 | 5 | 34 | 54 | 116 | 72 | 32 | 541 |
| 5 | 1 | .. | 12 | 20 | 7 | .. | 45 |
| 79 | 123 | 80 | 135 | 200 | 85 | 38 | 1,310 |
| 51 | 80 | 44 | 51 | 37 | 2 | 2 | 302 |
| .. | .. | .. | 1 | 1 | .. | .. | 3 |
| 4 | 11 | 16 | 10 | 4 | 7 | .. | 53 |
| 55 | 91 | 60 | 62 | 42 | 9 | 2 | 358 |
| .. | .. | .. | .. | .. | .. | .. | 16,600 |
| 4,800 | 7,100 | 3,900 | 100 | .. | .. | .. | 45,000 |
| 1,700 | 100 | .. | .. | .. | .. | .. | 1,800 |
| 1,900 | 12,400 | 2,000 | .. | .. | .. | .. | 16,800 |
| 45,700 | 41,600 | 24,700 | .. | .. | .. | .. | 228,500 |
| 15,900 | 21,600 | 10,200 | 8,200 | .. | .. | .. | 56,200 |
| 1,800 | .. | .. | .. | .. | .. | .. | 1,800 |
| 292,200 | 489,600 | 289,800 | 426,000 | 171,300 | 106,500 | .. | 2,458,500 |
| 93,100 | 113,000 | 169,500 | 91,700 | 125,300 | 81,200 | .. | 1,361,400 |
| 8,100 | 15,400 | .. | .. | 6,000 | .. | .. | 33,400 |
| .. | .. | .. | .. | 100 | .. | .. | 100 |
| 3,700 | 300 | 300 | .. | .. | .. | .. | 9,400 |
| 800 | 3,700 | .. | .. | .. | .. | .. | 4,500 |
| 26,200 | 271,300 | 134,100 | 192,100 | 110,100 | 38,000 | .. | 970,800 |
| 300 | 22,300 | 6,700 | 38,600 | 50,500 | 24,100 | .. | 239,600 |
| 17,800 | 20,200 | 900 | 9,000 | 5,400 | 4,800 | .. | 75,500 |
| .. | .. | .. | .. | 100 | 300 | .. | 400 |
| 1,300 | 71,300 | 300 | .. | .. | .. | .. | 146,700[1] |
| 500 | .. | .. | .. | 7,100 | .. | .. | 7,600[1] |
| 46,300 | 16,800 | 8,300 | 4,600 | .. | .. | .. | 78,700 |
| .. | .. | .. | 400 | .. | .. | .. | 400 |
| .. | .. | .. | 500 | 1,500 | 10,400 | .. | 12,400 |
| 562,100 | 1,106,700 | 650,700 | 771,200 | 477,400 | 265,300 | | 5,766,100 |

:ll filled for anti-aircraft purposes.

## APPENDIX XXXIX

### NUMBER OF BRITISH ANTI-AIRCRAFT GUNS ON THE WESTERN FRONT INCLUDING INDEPENDENT FORCE, JULY AND NOVEMBER 1918

*July 1918*

| | |
|---|---|
| 10 | 12-pdrs. |
| 203 | 13-pdrs. |
| 76 | 3-inch |
| 4 | 2-pdr. Vickers pom-pom |

293—*147 sections. Including Independent Force.*

*Armistice*

| | |
|---|---|
| 10 | 12-pdrs. |
| 244 | 13-pdrs. |
| 106 | 3-inch—2 R.M.A. |
| 4 | 2-pdr. pom-pom |

364—*182 sections. Including Independent Force.*


## APPENDIX XLII

### LENGTH OF FRONT HELD BY BRITISH IN FRANCE

*Various dates, 1917 and 1918*

| | |
|---|---|
| On 1st July 1917 | 80 miles |
| In December 1917 (after close of Battles of Ypres 1917 and Cambrai) | 85 ,, |
| In 1918 about 21st March, after the extension of the British front to Barisis | 100 ,, |
| In May 1918, after the close of the German offensives on the Somme and Lys | 90 ,, |
| On 8th August 1918 | 90 ,, |
| On the 11th November 1918 | 60 ,, |

## APPENDIX XLIII

### HOSTILE BOMBING ACTIVITY ON BRITISH FRONT IN FRANCE, MAY–OCTOBER 1918

| Month. | No. of enemy aeroplanes reported. | No. of bombs dropped. | No. of bombs in L. of C. area. | No. of bombs in Army Areas. | No. of Bombing Nights. | Average no. of bombs per Bombing Night. | Casualties. | | Enemy aircraft damaged or destroyed. | | | |
|---|---|---|---|---|---|---|---|---|---|---|---|---|
| | | | | | | | Killed. | Wounded. | Destroyed by 151 Squadron. | Gun-fire. | | S.A. Fire destroyed. |
| | | | | | | | | | | Damaged. | Destroyed. | |
| May | 896 | 3,720 | 2,028 | 1,692 | 22 | 169 | 530 | 1,353 | .. | 9 | 19 | 8 |
| June | 572 | 2,313 | 1,226 | 1,087 | 25 | 92 | 103 | 406 | .. | 13 | 6 | 3 |
| July | 564 | 2,177 | 590 | 1,587 | 22 | 99 | 179 | 424 | 1 | 5 | 12 | 1 |
| August | 854 | 3,287 | 745 | 2,542 | 21 | 156 | 179 | 501 | 5 | 10 | 16 | 2 |
| September | 512 | 3,001 | 253 | 2,748 | 21 | 143 | 120 | 259 | 12 | 19 | 12 | Nil |
| October | 316 | 1,587 | Nil | 1,587 | 16 | 99 | 41 | 61 | 2 | 6 | 17 | 1 |

# APPENDIX XLIV

## SUMMARY STATISTICS OF GERMAN AIR RAIDS ON GREAT BRITAIN 1914-18

### (A) GREAT BRITAIN (including M.P.D. and County of London)

| Type of raid. | No. of raids. | Bombs dropped. | Weight. tons. | Weight. cwt. | Weight. lb. | Casualties. Killed. | Casualties. Injured. | Estimated monetary damage. | Average. Weight of bomb dropped | Average. Casualty per bomb. | Average. Monetary damage per bomb |
|---|---|---|---|---|---|---|---|---|---|---|---|
| Airship . . | 51 | 5,806 | 196 | 8 | 64 | 557 | 1,358 | £1,527,585 | 75 lb. | 0·33 | £263 |
| Aeroplane . . | 52 | 2,772 | 73 | 11 | 69 | 857 | 2,058 | £1,434,526 | 59 lb. | 1·05 | £520 |
| TOTALS | 103 | 8,578 | 270 | 0 | 21 | 1,414 | 3,416 | £2,962,111 | .. | 0·56 | £345 |

### (B) METROPOLITAN POLICE DISTRICT (M.P.D.) (includes County of London)
*(covers an area within a radius of approximately 15 miles from Charing Cross)*

| Airship . . | 12 | 746 | 24 | 13 | 82 | 183 | 516 | no figures available | 75 lb. | 0·93 | .. |
|---|---|---|---|---|---|---|---|---|---|---|---|
| Aeroplane . . | 19 | 1,078 | 31 | 10 | 82 | 487 | 1,444 | ,, | 66 lb. | 1·80 | .. |
| TOTAL | 31 | 1,824 | 66 | 4 | 52 | 670 | 1,960 | .. | .. | 1·44 | .. |

### (C) COUNTY OF LONDON (includes City of London and covers area within a radius of approximately 6 miles from Charing Cross)

| Airship . . | 7 | 398 | 11 | 16 | 25 | 153 | 406 | £850,109 | 66 lb. | 1·40 | £2,136 |
|---|---|---|---|---|---|---|---|---|---|---|---|
| Aeroplane . . | 18 | 843 | 23 | 12 | 102 | 441 | 1,302 | £1,193,090 | 62 lb. | 2·06 | £1,415 |
| TOTAL | 25 | 1,241 | 35 | 9 | 15 | 594 | 1,708 | £2,043,199 | .. | 1·85 | £1,646 |

## APPENDIX XLV

### ANTI-AIRCRAFT DEFENCES IN GREAT BRITAIN

SCHEDULE OF TYPES, DISPOSITION, AND STRENGTHS OF AIRCRAFT, GUNS, HEIGHTFINDERS, SEARCHLIGHTS, SOUND-LOCATORS, AND STRENGTH OF PERSONNEL, 10th JUNE 1918

| | Guns | | | | | | | | | | | | | | | Heightfinders. | | | | | |
|---|---|---|---|---|---|---|---|---|---|---|---|---|---|---|---|---|---|---|---|---|---|
| | Approved armament. | | | | | | | Ready for action. | | | | | | | Approved. | | | Ready for action. | | |
| | 13-pdr. 9-cwt. | 4-in. Q.F. | 18-pdr. | 3-in. 20-cwt. | 75-mm. | 12-pdr. 12-cwt. | Total. | 13-pdr. 9-cwt. | 4-in. Q.F. | 18-pdr. | 3-in. 20-cwt. | 75-mm. | 12-pdr. 12-cwt. | Total. | Electrical. | Telephonic. | Total. | Electrical. | Telephonic. | Total. |
| Cromarty and Inverness | .. | .. | .. | .. | .. | 6 | 6 | .. | .. | 2 | .. | .. | 4 | 6 | 3 | .. | 3 | 2 | .. | 2 |
| Tay—Aberdeen—Longside | .. | .. | .. | .. | .. | 5 | 5 | .. | .. | 5 | .. | .. | .. | 5 | .. | 5 | 5 | .. | 5 | 5 |
| Forth | .. | .. | .. | 17 | 8 | .. | 25 | .. | .. | 8 | 9 | .. | .. | 25* | 15 | 2 | 17 | 12 | 3 | 15 |
| Tyne | .. | .. | .. | 20 | .. | 4 | 24 | .. | .. | 5 | 19 | 8 | .. | 24 | 15 | .. | 15 | 10 | 4 | 14 |
| Tees | .. | 1 | .. | 13 | .. | .. | 14 | .. | .. | .. | 14 | .. | .. | 14 | 11 | .. | 11 | 10 | 1 | 11 |
| Leeds | .. | .. | .. | 30 | .. | .. | 30 | .. | .. | .. | 25 | 5 | .. | 30 | 22 | 1 | 23 | 21 | 2 | 23 |
| Humber | .. | .. | .. | 18 | .. | .. | 18 | .. | .. | .. | 18 | .. | .. | 18 | 13 | 2 | 15 | 9 | 4 | 13 |
| Manchester | .. | .. | .. | 14 | .. | 8 | 22 | .. | .. | 2 | .. | .. | 20 | 22 | 13 | 6 | 19 | 11 | 7 | 18 |
| Nottingham | .. | .. | .. | 15 | .. | .. | 15 | .. | .. | .. | 5 | .. | 10 | 15 | 9 | 5 | 14 | 5 | 8 | 13 |
| Birmingham | .. | .. | .. | .. | 20 | .. | 20 | .. | .. | .. | .. | 14 | .. | 14 | 9 | 1 | 10 | 7 | 3 | 10 |
| TOTAL | .. | 1 | .. | 127 | 28 | 23 | 179 | .. | .. | 22 | 90 | 27 | 34 | 173 | 110 | 22 | 132 | 87 | 37 | 124 |

*Northern air defences (permanent).*

# ANTI-AIRCRAFT DEFENCES

*London air defence area (permanent).*

| | Guns — Approved armament | | | | | | | Guns — Ready for action | | | | | | | Heightfinders — Approved | | | Heightfinders — Ready for action | | |
|---|---|---|---|---|---|---|---|---|---|---|---|---|---|---|---|---|---|---|---|---|
| | 13-pdr. 9-cwt. | 4-in. Q.F. | 18-pdr. | 3-in. 20-cwt. | 75-mm. | 12-pdr. 12-cwt. | Total | 13-pdr. 9-cwt. | 4-in. Q.F. | 18-pdr. | 3-in. 20-cwt. | 75-mm. | 12-pdr. 12-cwt. | Total | Electrical | Telephonic | Total | Electrical | Telephonic | Total |
| Central London | .. | .. | .. | 22 | .. | .. | 22 | .. | .. | .. | 22 | .. | .. | 22 | 15 | .. | 15 | 13 | 2 | 15 |
| West London | .. | .. | .. | 16 | .. | .. | 16 | .. | .. | .. | 16 | .. | .. | 16 | 12 | .. | 12 | 9 | 6 | 15 |
| Epping | .. | .. | .. | 43 | .. | .. | 43 | .. | .. | .. | 43 | .. | .. | 43 | 11 | .. | 11 | 4 | 7 | 11 |
| St. Albans | .. | .. | 31 | 4 | .. | .. | 35 | .. | .. | 21 | 3 | .. | .. | 24 | 10 | 1 | 11 | .. | 5 | 5 |
| Staines | .. | .. | 20 | 3 | .. | .. | 23 | .. | .. | 19 | .. | .. | .. | 19 | 10 | .. | 10 | 1 | 6 | 7 |
| Redhill | .. | .. | 26 | .. | .. | .. | 26 | .. | .. | 13 | .. | .. | .. | 13 | 7 | .. | 7 | .. | 3 | 3 |
| Harwich | .. | .. | .. | 14 | .. | .. | 14 | .. | .. | 2 | 12 | .. | .. | 14 | 6 | 3 | 9 | 5 | 4 | 9 |
| Thames and Medway | .. | .. | .. | 38 | .. | .. | 38 | .. | .. | .. | 38 | .. | .. | 38 | 17 | .. | 17 | 10 | 4 | 14 |
| Dover | .. | .. | .. | 20 | .. | .. | 20 | .. | .. | .. | 20 | .. | .. | 20 | 11 | .. | 11 | 11 | .. | 11 |
| Portsmouth | .. | .. | 10 | 10 | .. | .. | 20 | .. | .. | 11 | 4 | .. | .. | 15 | 8 | 1 | 9 | 8 | 1 | 9 |
| Total | .. | .. | 87 | 170 | .. | .. | 257 | .. | .. | 66 | 158 | .. | .. | 224 | 107 | 5 | 112 | 61 | 38 | 99 |

# IN GREAT BRITAIN

| | Guns — Approved armament | | | | | | | Guns — Ready for action | | | | | | | Heightfinders — Approved | | | Heightfinders — Ready for action | | |
|---|---|---|---|---|---|---|---|---|---|---|---|---|---|---|---|---|---|---|---|---|
| | 13-pdr. 9-cwt. | 4-in. Q.F. | 18-pdr. | 3-in. 20-cwt. | 75-mm. | 12-pdr. 12-cwt. | Total. | 13-pdr. 9-cwt. | 4-in. Q.F. | 18-pdr. | 3-in. 20-cwt. | 75-mm. | 12-pdr. 12-cwt. | Total. | Electrical | Telephonic | Total | Electrical | Telephonic | Total |
| *Mobiles (London air defence area).* | | | | | | | | | | | | | | | | | | | | |
| No. 1 Mobile Brigade | 18 | .. | .. | .. | .. | .. | 18 | 18 | .. | .. | .. | .. | .. | 18 | .. | 9 | 9 | .. | 9 | 9 |
| No. 2 Mobile Brigade | 18 | .. | .. | .. | .. | .. | 18 | 18 | .. | .. | .. | .. | .. | 18 | .. | 9 | 9 | .. | 9 | 9 |
| 3 Batteries, Nos. 8, 9, and 10 | 18 | .. | .. | .. | .. | .. | 18 | 18 | .. | .. | .. | .. | .. | 18 | .. | 9 | 9 | .. | 9 | 9 |
| Total | 54 | .. | .. | .. | .. | .. | 54 | 54 | .. | .. | .. | .. | .. | 54 | .. | 27 | 27 | .. | 27 | 27 |
| *Training guns.* | | | | | | | | | | | | | | | | | | | | |
| School of Gunnery | 3 | 1 | 2 | 4 | 1 | .. | 11 | 3 | 1 | 1 | 4 | 1 | .. | 10 | 1 | 2 | 3 | 1 | 2 | 3 |
| Reserve Brigade | 4 | .. | 1 | 3 | .. | .. | 8 | 4 | .. | 1 | 3 | .. | .. | 8 | 1 | 2 | 3 | 1 | 2 | 3 |
| Leasowe | .. | .. | .. | .. | .. | .. | .. | .. | .. | .. | .. | .. | .. | .. | .. | 1 | 1 | .. | 1 | 1 |
| Ordnance College | .. | 1 | .. | .. | 1 | .. | .. | .. | 1 | .. | .. | 1 | .. | .. | 1 | .. | 1 | 1 | .. | 1 |
| Total | 7 | 2 | 3 | 7 | 1 | .. | 19 | 7 | 1 | 2 | 7 | 1 | .. | 18 | 3 | 5 | 8 | 3 | 5 | 8 |
| Grand Total U.K. | 61 | 2 | 90 | 304 | 29 | 23 | 509 | 61 | 1 | 90 | 255 | 28 | 34 | 469 | 220 | 59 | 279 | 151 | 107 | 258 |
| Unallotted | .. | .. | .. | 1 | .. | 14 | 15 | .. | .. | .. | .. | .. | .. | .. | .. | .. | .. | .. | .. | .. |

# ANTI-AIRCRAFT DEFENCES

NOTES

(1) The following A.A. guns had been mounted by the Admiralty on shore and in ships stationed permanently or semi-permanently in rivers and estuaries. For operations they came under the orders of A.A. Defence Commanders as follows:

2–3-in. 20-cwt. Harwich  
2–3-in. 20-cwt. Lowestoft } A.A.D.C., Harwich.  
1–3-in. 20-cwt. Yarmouth  
6–3-in. 20-cwt. Chatham } A.A.D.C., Thames and Medway.  
4–3-in. 20-cwt. Sheerness  
2–3-in. 20-cwt. Dover. A.A.D.C., Dover.  
2–3-in. 20-cwt. Kinghorn Ness. A.A.D.C., Forth.  
2–3-in. 20-cwt. Teesmouth. A.A.D.C., Tees.  
2–3-in. 20-cwt. Humber } A.A.D.C., Humber.  
2–12-pdr. 12-cwt. Immingham  

Total 25

(2) *Mobile Guns.*

No. 1 Mobile Brigade, together with Nos. 9 and 10 Mobile Batteries attached, were in KENT distributed on a line Herne Bay–Ashford.

No. 2 Mobile Brigade was in ESSEX distributed on a line Brightlingsea–Great Wakering.

No. 8 Mobile Battery was on Mersea Island and was under the orders of the A.A. Defence Commander, Harwich.

All the above mobile guns were in the LONDON Air Defence Area.

(3) *4-in. A.A. guns.*

No definite allotment had been made up to this date owing to the uncertainty in the rate of supply; the policy, however, was to replace 3-in. 20-cwt. guns in the North and Midlands at places which were particularly liable to Zeppelin raids.

(4) *12-pdr. 12-cwt. guns shown as unallotted in the 'Project'.*

These guns were in use as temporary equipment at various sites, but the intention was to return them as and when the full equipment of guns was received.

(5) *Replacement of 3-in. 20-cwt. by 18-pdrs. in the London barrier.*

It had been found necessary to move all 18-pdrs. to the LONDON barrier, as this gun was capable of dealing with aeroplanes, and to move 3-in. 20-cwt. guns to the North and Midlands because these were the only guns available capable of dealing with high-flying Zeppelins.

# SEARCHLIGHTS, ETC.

| | 120 cm. or larger (no. approved). | 150 cm. | 120 cm. | Single 90 cm. | Pair 60 cm. | Twin 60 cm. | Single 60 cm. | Spare mobile. | Mobile fighting. | Total. | Approved. | Ready for action. |
|---|---|---|---|---|---|---|---|---|---|---|---|---|
| | | | | Searchlights ready for action. | | | | | | | Sound Locators. | |
| | | | | *Northern Air Defences (Permanent).* | | | | | | | | |
| Cromarty and Inverness | 6 | .. | .. | .. | 1 | .. | 2 | .. | .. | 3 | 6 | .. |
| Tay—Aberdeen—Longside | 5 | .. | .. | .. | .. | .. | 5 | .. | .. | 5 | 5 | .. |
| Forth | 36 | .. | .. | .. | 3 | .. | 28 | 2 | .. | 33 | 36 | .. |
| Tyne | 36 | .. | .. | .. | 4 | .. | 32 | .. | .. | 36 | 36 | .. |
| Tees | 29 | .. | .. | .. | 4 | .. | 25 | .. | .. | 29 | 29 | .. |
| Leeds | 45 | .. | .. | .. | 6 | .. | 39 | .. | .. | 45 | 45 | .. |
| Humber | 34 | .. | .. | .. | 5 | .. | 29 | .. | .. | 34 | 34 | .. |
| Manchester | 24 | .. | .. | .. | 3 | .. | 21 | .. | .. | 24 | 24 | .. |
| Nottingham | 36 | .. | .. | .. | .. | .. | 35 | 1 | .. | 36 | 36 | .. |
| Birmingham | 13 | .. | .. | .. | 1 | 1 | 12 | .. | .. | 14 | 13 | .. |
| TOTAL | 264 | .. | .. | .. | 27 | 1 | 228 | 3 | .. | 259 | 264 | .. |
| | | | | *London Air Defence Area (Permanent).* | | | | | | | | |
| Central London | 40 | 2 | 16 | .. | .. | 8 | 14 | .. | .. | 40 | 40 | 4 |
| East London | 52 | 5 | 7 | .. | .. | .. | 39 | .. | .. | 51 | 52 | .. |
| West London | 30 | .. | 4 | .. | 1 | 4 | 21 | .. | .. | 30 | 30 | .. |
| Epping | 43 | .. | .. | 4 | 1 | .. | 37 | 1 | .. | 43 | 43 | .. |
| St. Albans | 22 | .. | .. | .. | .. | .. | 17 | .. | 1 | 18 | 22 | .. |
| Staines | 19 | .. | .. | .. | .. | .. | 9 | .. | .. | 9 | 19 | .. |
| Redhill | 13 | .. | .. | .. | .. | .. | 11 | .. | .. | 11 | 13 | .. |
| Harwich | 50 | .. | .. | 3 | .. | .. | 40 | 6 | 1 | 50 | 50 | .. |
| Thames and Medway | 28 | .. | .. | 4 | 1 | .. | 23 | .. | .. | 28 | 28 | .. |
| Dover | 38 | .. | 1 | 6 | .. | .. | 22 | .. | 10 | 39 | 38 | .. |
| Portsmouth | 23 | .. | .. | .. | .. | 3 | 13 | 1 | .. | 17 | 23 | .. |
| TOTAL | 358 | 7 | 28 | 17 | 3 | 15 | 246 | 8 | 12 | 336 | 358 | 4 |
| | | | | *Mobiles (London Air Defence Area).* | | | | | | | | |
| No. 1 Mob. Bde. | 9 | .. | .. | .. | .. | .. | .. | .. | 9 | 9 | 9 | .. |
| No. 2 Mob. Bde. | 9 | .. | .. | .. | .. | .. | .. | .. | 8 | 8 | 9 | .. |
| 3 Batteries, Nos. 8, 9, and 10 | 9 | .. | .. | .. | .. | .. | .. | 2 | .. | 2 | 9 | .. |
| TOTAL | 27 | .. | .. | .. | .. | .. | .. | 2 | 17 | 19 | 27 | .. |

# ANTI-AIRCRAFT DEFENCES

| | 120 cm. or larger (no. approved) | Searchlights ready for action. | | | | | | | | Sound Locators. | |
|---|---|---|---|---|---|---|---|---|---|---|---|
| | | 150 cm. | 120 cm. | Single 90 cm. | Pair 60 cm. | Twin 60 cm. | Single 60 cm. | Spare mobile. | Mobile fighting. | Total. | Approved. | Ready for action. |

*Searchlights and Sound Locators at Training Establishments.*

| | | | | | | | | | | | | |
|---|---|---|---|---|---|---|---|---|---|---|---|---|
| School of Gunnery | 2 | .. | .. | 1 | .. | .. | 1 | .. | .. | 2 | 2 | .. |
| Stokes Bay | 6 | .. | .. | .. | .. | .. | 6 | .. | .. | 6 | 13 | 6 |
| TOTAL | 8 | .. | .. | 1 | .. | .. | 7 | .. | .. | 8 | 15 | 6 |
| Grand Total U.K. | 657 | 7 | 28 | 18 | 30 | 16 | 481 | 13 | 29 | 622 | 664 | 10 |
| Unallotted | 11 | .. | .. | .. | .. | .. | .. | .. | .. | .. | 11 | .. |
| TOTAL | 668 | .. | .. | .. | .. | .. | .. | .. | .. | .. | 675 | .. |

*Note.* 240 of the 668 lights approved for Home Defence were to be of 120 cm. size or larger. A proposal, however, was under consideration that all searchlights should be of 120 cm. size or larger, and that the total number allotted to Home Defence should be increased to 808.

## ESTABLISHMENT AND FIGHTING STRENGTH OF 6th BRIGADE, ROYAL AIR FORCE

### AIRCRAFT

| | No. of Squadron. | Head-quarters. | Establishment. | No. of efficient aircraft on 8th June 1918. |
|---|---|---|---|---|
| Northern Air Defences | 77 | Penston | 22 Bristol Fighters | Nil |
| | 36 | Hylton | 20 ,, ,, | 6 Bristol Fighters |
| | 76 | Ripon | 24 ,, ,, | 1 ,, ,, |
| | 33 | Kirton Lindsey | 22 ,, ,, | 6 ,, ,, |
| | 38 | Buckminster | 24 ,, ,, | Nil (overseas) |
| | 51 | Marham | 24 ,, ,, | Nil |
| | 75 | Elmswell | 22 ,, ,, | Nil |
| | 37 | Stow Maries | 24 S.E.5 'Viper' | 6 S.E.5a } 7 Sopwith 'Camels' } |
| London Air Defence Area | 61 | Rochford | 24 ,, ,, | 10 S.E.5a |
| | 39 | North Weald Bassett | 24 Bristol Fighters | 21 Bristol Fighters } 2 Martinsyde ,, } |
| | 44 | Hainault Farm | 24 Sopwith 'Camels' | 19 Sopwith 'Camels' |
| | 141 | Biggin Hill | 24 Bristol Fighters | 22 Bristol Fighters |
| | 78 | Sutton's Farm | 24 Sopwith 'Camels' | 20 Sopwith 'Camels' |
| | School of Aerial Co-operation | Gosport | 6 S.E.5 'Viper' | Nil |
| | 143 | Detling | 22 ,, ,, | 14 S.E.5a |
| | 112 | Throwley | 24 Sopwith 'Camels' | 20 Sopwith 'Camels' |
| | 50 | Bekesbourne | 22 S.E.5 'Viper' | 12 S.E.5a |
| TOTALS | | | 376 aeroplanes | 166 aeroplanes |

# ESTABLISHMENTS

## DISTRIBUTION

|  | Establishment. | On charge. | Deficient. |
|---|---|---|---|
| Northern Air Defences | 112 | 13 | 99 |
| London Air Defence Area | 264 | 153 | 111 |
| TOTALS | 376 | 166 | 210 |

*Notes:*

(1) There were, in addition, a number of aeroplanes which were used for training purposes and for the communication of information regarding the enemy, by W/T. These could not be classed as efficient fighters.

(2) In addition to the efficient aircraft above mentioned there were approximately 30 efficient aeroplanes available from the Training Squadrons, Acceptance Parks, and Experimental Stations for DAY RAIDS ONLY, and were given a certain role of action.

## PERSONNEL

|  | War Establishment, H.Q., Brigade, Wings, and Fighting Squadrons. | | | Fighting strength on 8th June 1918. | | |
|---|---|---|---|---|---|---|
|  | Officers. | Other Ranks. | Women. | Officers. | Other Ranks. | Women. |
| 6th Brigade | 576 | 3,548 | 448 | 660 | 3,639 | 315 |

|  | War Establishment. | | Fighting strength on 8th June 1918. | |
|---|---|---|---|---|
|  | Officers. | Other Ranks. | Officers. | Other Ranks. |
| No. 7 Balloon Wing, R.A.F. | 120 | 3,475 | 82 | 2,573 |

## ESTABLISHMENTS AND FIGHTING STRENGTHS OF ANTI-AIRCRAFT UNITS R.G.A. AND R.E.

|  | War Establishment. | | 1st January 1918 Strength. | June 1918 Strength. |
|---|---|---|---|---|
|  | Dec. 1917. | June 1918. | | |
| **R.G.A.** | | | | |
| Officers | 639 | 639 | 537 | 603 |
| Other ranks | 8,436 | 8,436 | 7,620 | 2,888 |
| **R.E.** | | | | |
| Officers | 108 | 129* | 138 | 127 |
| Other ranks | 4,201 | 4,620* | 3,527 | 2,518 |
| TOTAL | 13,384 | 13,824 | 11,822 | 6,136 |

\* Includes establishment for Nos. 13, 14, 15, 16, 17 A.A. Coys. R.E.

# APPENDIX XLVI

## STRENGTH IN PERSONNEL OF ROYAL AIR FORCE IN VARIOUS THEATRES OF WAR AT 31st OCTOBER 1918

| Theatre. | Officers. Combatant. | Officers. Non-combatant. | Other ranks. |
|---|---|---|---|
| Dover–Dunkirk, France, Independent Force | 3,742 | 1,440 | 48,893 |
| Italy | 138 | 48 | 1,480 |
| Mediterranean | 460 | 171 | 5,710 |
| Palestine | 158 | 55 | 1,579 |
| Mesopotamia | 89 | 30 | 1,039 |
| Salonika | 108 | 29 | 1,189 |
| India and Aden | 46 | 26 | 535 |
| Grand Fleet and Orkneys and Shetlands | 345 | 54 | 3,364 |
| Egypt (Training) | 224 | 254 | 6,283 |
| Canada " | 399 | 197 | 10,892 |
| Russia | 30 | .. | 250 |
| | 5,739 | 2,304 | 81,214 |

| | Officers. | Other ranks (including cadets under instruction). |
|---|---|---|
| Home Areas | 11,389 | 181,305 |
| American Mission | 29 | 67 |
| Hospital | 300 | .. |
| Vendome | 49 | 1,256 |
| | 11,767 | 182,628 |

In addition to the above figures there were between 7,000 and 8,000 officers attached to various home units for instruction.

The numbers of pilots and observer pupils under instruction at home were as follows:

| | Officers. | Cadets. | Other ranks. |
|---|---|---|---|
| Non-flying Units | 1,939 | 10,263 | 164 |
| Flying Training Units | 5,032 | 3,153 | 1,418 |
| Total Pilot Pupils | 6,971 | 13,416 | 1,582 |
| Total Observer Pupils | 552 | 1,672 | 11 |

There were approximately 5,800 officers and 67,000 other ranks on the strength of Training Depot Stations, Schools, and other Training Units at Home, exclusive of those under instruction. In this total of officers are included 1,601 flying instructors.

Of the total personnel in home areas there were 1,710 combatant officers, 629 non-combatant officers, and about 22,500 other ranks employed in Home Defence and Coastal Operations.

The numbers employed on administrative work at head-quarters, in administrative units other than Schools, and at Depots, were about 2,896 officers and 71,100 other ranks.

In units preparing for overseas there were 354 officers and 3,200 other ranks.

Of the strength of other ranks at home, about 1,000 were in hospital or awaiting discharge.

*Agreement of Personnel in Home Areas with Figure Previously Shown*

|  | Officers. | Other ranks. |
|---|---|---|
| Home Defence and Coastal Operations | 2,339 | 22,500 |
| Cadets and N.C.O. Pupils | .. | 16,500 |
| Administrative Units and Depots | 2,896 | 71,100 |
| Training Depot Stations, &c. | 5,800 | 67,000 |
| Units preparing for Overseas | 354 | 3,200 |
| In Hospital, &c. | .. | 1,000 |
|  | 11,389 | 181,300 |

The casualties in combatant personnel during the month of OCTOBER were as follows:

|  | Officers. | Cadets. | N.C.O.s. |
|---|---|---|---|
| Home Training | 137 | 61 | 5 |
| Canada „ | 2 | 7 | .. |
| Egypt „ | 14 | 6 | 2 |
| Home Battle | 8 | .. | 24 |
| France „ | 545 | .. | 35 |
| Italy „ | 22 | .. | .. |
| Middle East | 2 | .. | .. |
|  | 730 | 74 | 66 |

In addition there were reported during OCTOBER 447 casualties (killed; died; wounded; injured; missing; &c.), amongst other ranks overseas.

# APPENDIX XVI
## ORGANIZATION OF R.A.F. MIDDLE EAST 30TH SEPTEMBER 1918

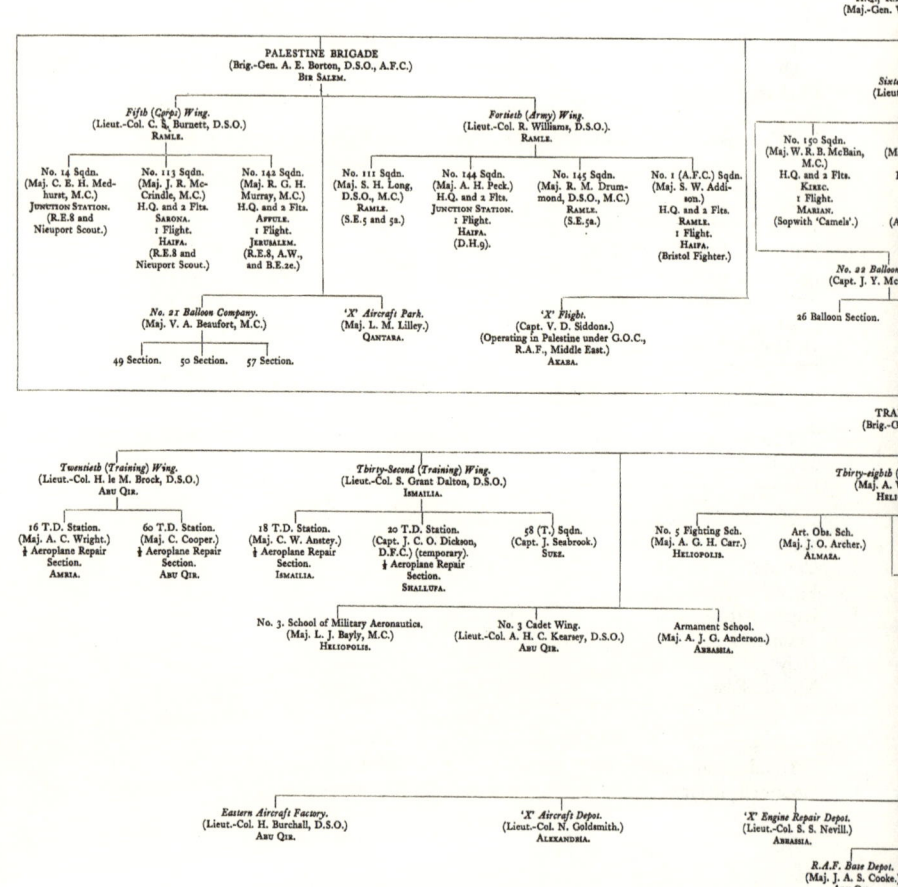

**IDDLE EAST**
(. Salmond, D.S.O.)
10.

**MESOPOTAMIA**
Thirty-first (H.Q.) Wing.
(Lieut.-Col. R. A. Bradley.)
BAGHDAD.

**INDIA**
Director of Aeronautics.
(Lieut.-Col. J. E. Tennant, D.S.O., M.C.)
SIMLA.

Fifty-Second (Corps) Wing.
(Lieut.-Col. J. R. C. Heathcote.)
PESHAWAR.

**NIKA**
orps) Wing.
J. E. Todd.)
SH.

Sqdn.
, Hodges,
C.)
d 1 Flt.
:OVO.
ghts.
REOJ.
E.12, and
-9.)

No. 47 Sqdn.
(Maj. F. A. Bates, M.C.)
H.Q. and 2 Flts.
HAJDARLI.
1 Flight.
YANESH.
(D.H.9 and A.W.)

No. 30 Sqdn.
(Maj. J. Everidge.)
H.Q. and 2 Flts.
KIFRI.
½ Flt. HAMADAN ⎫
½ Flt. ZINJAN  ⎬ Persia.
(R.E.8 and B.E.2c.)

No. 63 Sqdn.
(Maj. F. L. Robinson, M.C.)
TIKRIT.
(R.E.8.)

No. 72 Sqdn.
(Maj. O. T. Boyd, M.C.)
H.Q. and 2 Flts.
ZINJAN.
1 Flight.
SAMARRA.
(Martinsyde, Bristol Monoplane, Sopwith 'Camel'.)

No. 31 Sqdn.
(Capt. O. Hughes.)
H.Q. and 2 Flts.
RISALPUR.
1 Flight.
KHANPUR.
(B.E.2c and e.)

No. 114 Sqdn.
(Maj. D. E. Stodart, D.S.O., D.F.C.)
H.Q. and 1½ Flts.—QUETTA.
½ Flight—LAHORE.
1 Flight.
ADEN.
(B.E.2c and e.)

No. 23 Balloon Company.
(Maj. H. D. Jensen, M.C.)

51 Section.    52 Section.

Port Depot.
BOMBAY.

oon Section.

·y,
.C.)

Salonika Aircraft Park.
(Maj. C. H. A. Hirtzel.)
MIKRA BAY.

Army Aircraft Park.
(Maj. G. Somers-Clarke.)
BAGHDAD.

Aircraft Park.
(Maj. R. S. Rumbold.)
RISALPUR.

**BRIGADE**
W. Herbert.)
POLIS.

g) Wing.
ler.)

Sixty-ninth (Training) Wing.
(Lieut.-Col. E. W. Powell.)
EL RIMAL.

Flying Instructors Sch.
aj. B. P. H. de Roeper.)
KHANKA.

Sch. of Navgn. and Bomb Dropping.
(Maj. G. Merton, M.C.)
HELWAN.

17 T.D. Station
(Maj. C. W. Hyde.)
½ Aeroplane Repair Section.
ABU SUEIR.

19 T.D. Station.
(Maj. G. M. Croil.)
½ Aeroplane Repair Section.
EL RIMAL.

½ Aeroplane Repair Section.
HELIOPOLIS.

Sixty-fourth (Naval) Wing.
(Lieut.-Col. C. E. Risk.)
ALEXANDRIA.

Seaplane Refuelling Base and Depot.
(Maj. P. L. Holmes, D.S.C.)
PORT SAID.

Seaplane Station.
(Maj. E. J. Hodsoll.)
ALEXANDRIA.

No. 269 Sqdn. (Port Said Seaplane Sqdn.).
(Maj. P. L. Holmes, D.S.C.)
PORT SAID.

No. 2 Balloon Base.
(Maj. M. Lyon.)
ALEXANDRIA.

No. 22 Balloon Base.
(Capt. T. R. Spence.)
PORT SAID.

*Note.* R.A.F. Units working with the Navy in Egypt were administered by G.O.C., R.A.F., Middle East. Questions of policy and operations were referred by the latter to the G.O.C., Mediterranean District. Administrative returns of personnel and material were forwarded, through the G.O.C., Mediterranean District to the Air Ministry.

'X' Balloon Repair Depot.
(Lieut. W. Sutherland.)
ABBASSIA.

*Stores Distributing Park.*
ISMAILIA.

Civil Base Depot.
ABU QIR.

# APPENDIX XVII

## SUMMARY OF ANTI-SUBMARINE AIR-PATROLS FROM 1st MAY 1918 to 12th NOVEMBER 1918 (HOME WATERS)

| Type of aircraft. | Daily average of machines. | | | Average daily number of patrols. | Proportion of patrols. | | Total time on patrol for the period of six months. | Daily average patrol. | | Average patrol estimated on station strength. | | Percentage of patrols curtailed by | | | Hostile submarines. | | Hours of flight and mileage per submarine sighted. | Mines sighted. |
|---|---|---|---|---|---|---|---|---|---|---|---|---|---|---|---|---|---|---|
| | On station strength. A. (1) | Ready for service. B. (2) | Proportion. B/A × 100. (3) | C. (4) | On station strength. C/A × 100. (5) | Ready for service. C/B × 100. (6) | Hours. (7) | Hrs. (8) | Mins. (9) | Hrs. (10) | Mins. (11) | Engine trouble. (12) | Weather. (13) | Sighted. (14) | Attacked. (15) | (16) | (17) |
| *Aeroplanes:* | | | | | | | | | | | | % | % | | | | |
| Light Bombers | 41 | 28 | 68 | 12 | 29·4 | 42·8 | 4,080 | 1 | 42 | .. | 30 | 5·6 | 6·5 | 16 | 11 | 255 (19,125 miles) | .. |
| Heavy Bombers (incl. Blackburn 'Kangaroos') | 7·7 | 5·6 | 73 | 0·94 | 12·2 | 16·7 | 600 | 3 | 15 | .. | 23 | 11·3 | 20·4 | 12 | 11 | 50 (3,500 miles) | .. |
| Temporary types | 140 | 114 | 81·4 | 55 | 39·3 | 48·3 | 18,552 | 1 | 43 | .. | 40 | 3·4 | 8 | 38 | 22 | 485 (21,825 miles) | .. |
| Aeroplanes: Total | 189 | 147 | 78 | 68 | 36 | 45 | 23,232 | 1 | 44 | .. | 37 | 3·9 | 8 | 66 | 44 | 352 | .. |
| *Seaplanes:* | | | | | | | | | | | | | | | | | |
| Babies | 46 | 19 | 41 | 8·5 | 18 | 44 | 3,075 | 1 | 49 | .. | 20 | 6·4 | 8·5 | 13 | 10 | 236 (16,520 miles) | .. |
| Two-Seaters ('Short' Type) | 170 | 66 | 38 | 35 | 20 | 53 | 17,558 | 2 | 42 | .. | 31 | 8·2 | 10·1 | 33 | 25 | 530 (26,500 miles) | .. |
| Large Americas | 85 | 26 | 31 | 9 | 11·6 | 34·6 | 5,421 | 3 | 6 | .. | 19 | 8·12 | 11 | 28 | 18 | 196 (11,590 miles) | 8 |
| Seaplanes: Total | 301 | 111 | 37 | 52 | 17·2 | 46·7 | 26,054 | 2 | 32 | .. | 36 | 8·1 | 10 | 74 | 53 | 351 | 8 |
| *Airships:* | | | | | | | | | | | | | | | | | |
| 'S.S.' Type | 55 | 37 | 67 | 21 | 38 | 57 | 25,284 | 6 | 8 | 2 | 20 | 2·2 | 12·2 | 23 | 15 | 1,099 (43,960 miles) | 60 |
| Coastal and C. Star | 14·4 | 11·6 | 80 | 4·36 | 30 | 37 | 7,439 | 8 | 47 | 2 | 34 | 6·4 | 17·1 | 3 | 2 | 2,416 (96,640 miles) | 2 |
| North Sea | 4 | 2·75 | 69 | 0·059 | 1·5 | 2·15 | 1,256 | 10 | 49 | 1 | 37 | 2·59 | 18·1 | .. | .. | .. | .. |
| Rigid | 1·57 | 1·32 | 84 | 0·023 | 1·47 | 1·74 | 542 | 11 | 46 | 1 | 45 | .. | 21·8 | 1 | .. | 542 | .. |
| Airships: Total | 75 | 52 | 70 | 26 | 34·6 | 50 | 34,521 | 6 | 14 | 2 | 2 | 2·7 | 12·6 | 27 | 18 | 1,278 | 62 |

APPEN

TYPES OF AIRCRAFT 1914–18:

| Aeroplane. Make and type. | Duty. | No. of Seats. | Engine. | Normal B.H.P. and R.P.M. at ground level. | Speed. Miles per hour at. 6,500 ft. | 10,000 ft. | 15,000 ft. | 16,500 ft. | Time in 6,500 ft. Time. | Rate. |
|---|---|---|---|---|---|---|---|---|---|---|
| Armstrong-Whitworth (Tractor-biplane) | Training | 2 | 90-h.p. R.A.F. 1a | 100 (at 1,800) | .. | 81 (at 8,000) | .. | .. | 26·5 | 200 |
|  | Corps Reconnaissance | 2 | 120-h.p. Beardmore | 133 (at 1,200) | .. | 83·5 (at 8,000) | .. | .. | 19 | 280 |
|  |  |  | 160-h.p. Beardmore | 170 (at 1,400) | 95 | 88 | .. | .. | 15·4 | 330 |
| Avro (Tractor-biplane) | General Purpose (179) | 2 | 80-h.p. Gnome | 84 (at 1,200) | 62 | .. | .. | .. | 25 | .. |
|  | Training (504 K.) | 2 | 100-h.p. Monosoupape | 105 (at 1,200) | 82 | 75 | .. | .. | .. | .. |
|  |  |  | 110-h.p. Le Rhône | 127 (at 1,250) |  |  |  |  |  |  |
| B.E.2 (Tractor-biplane) | General Purpose | 2 | 70-h.p. Renault | 80 (at 1,800) | 65 | .. | .. | .. | 9 (to 3,000) | .. |
| B.E.2c (Tractor-biplane) | Corps Reconnaissance | 2 | 90-h.p. R.A.F. 1a | 100 (at 1,800) | 72 | 69 | .. | .. | 20 | 300 |
| B.E.2e (Tractor-biplane) | Corps Reconnaissance | 2 | 90-h.p. R.A.F. 1a | 100 (at 1,800) | 82 | 75 | .. | .. | 24 | 182 |
| B.E.8 (Tractor-biplane) | General Purpose | 2 | 80-h.p. Gnome | 84 (at 1,200) | 70 (ground level) | .. | .. | .. | 10·5 (to 3,000) | .. |
| B.E.12 (Tractor-biplane) | Fighter | 1 | 150-h.p. R.A.F. 4a | 160 (at 1,800) | 97 | 91 | .. | .. | 9·5 | 490 |
| B.E.12b (Tractor-biplane) | Fighter | 1 | 200-h.p. Hispano-Suiza | 210 (at 2,000) | (No performance figures available) | | | | | |
| Blackburn-Kangaroo (Tractor-biplane) | Bomber | 3 | 2–250-h.p. Rolls-Royce (Falcon) | 2—253 (at 2,000) | 98 (Performance | 86 without bombs—4 | .. 230 lb.) | .. | 18·2 | 250 |
| Bleriot (Tractor-Monoplane) | General Purpose | 2 | 80-h.p. Gnome | 84 (at 1,200) | 66 (ground level) | .. | .. | .. | 14 (to 3,000) | .. |
| Bristol Scout (Tractor-biplane | Fighter | 1 | 1. 80-h.p. Gnome 2. 80-h.p. Le Rhône | 84 (at 1,200) | 89 | 86·5 | .. | .. | 10·8 | 385 |
| Bristol Fighter (Tractor-biplane | Fighter Reconnaissance | 2 | 1. 250-h.p. Rolls-Royce (Falcon) | 253 (at 2,000) | 119 | 113 | 105 | .. | 6·5 | 830 |
| Bristol Fighter (Tractor-biplane) | Corps Reconnaissance | 2 | 2. 200-h.p. Hispano-Suiza | 210 (at 2,000) | .. | 105 | 97·5 | .. | 8·7 | 620 |
| Bristol Monoplane (Tractor-monoplane) | Fighter | 1 | 110-h.p. Le Rhône | 136 (at 1,400) | .. | 111·5 | 104 | .. | 5·5 | 885 |
| Caudron (Tractor biplane) | General Purpose | 2 | 80-h.p. Gnome or Le Rhône | 84 (at 1,200) | 66–71 (ground level) | .. | .. | .. | 20 | .. |
| Caudron Twin (Tractor-biplane) | Day Bomber | 2 | 2–80-h.p. Le Rhône or 100-h.p. Anzani | 168–190 h.p. (at 1,200) | 82 | 80 | .. | .. | 17 | .. |
| Curtiss (Tractor-biplane) | Training | 2 | 90-h.p. Curtiss Ox | 93 (at 1,400) | 70 | .. | .. | .. | 10 (to 3,000) | .. |
| de Havilland 1a (D.H.1a: Trac- | Fighter Reconnaissance | 2 | 120-h.p. Beardmore | 133 (at 1,200) | 90 | 86 | .. | .. | 14·8 | 340 |

# XXVII

## CAL DATA, TABLE A. AEROPLANES

| Climb. and rate in feet per min. | | | | | Air Endurance. Hours. | Service Ceiling. Feet. | Weight, lb. | | | | Dimensions. | | | | Period of use. Year. |
|---|---|---|---|---|---|---|---|---|---|---|---|---|---|---|---|
| 00 ft. Rate. | 15,000 ft. Time. | 15,000 ft. Rate. | 16,500 ft. Time. | 16,500 ft. Rate. | | | Gross. | Empty. | Fuel and Oil. | Military load and crew. | Surface Sq. ft. | Span. Ft. | Length. Ft. | Height. Ft. | |
| 143 | .. | .. | .. | .. | 3 | 12,000 | 2,056 | 1,386 | 230 | 440 | .. | 40' 5" | 29' 7" | 10' 9" | 1916–17 |
| 180 | .. | .. | .. | .. | 3 | 12,000 | 2,447 | 1,682 | 340 | 427 | .. | 42' 5" | 29' 9" | 10' 11" | 1917–18 |
| 240 | .. | .. | .. | .. | 3 | 13,000 | 2,811 | 1,916 | 402 | 493 | 504 | 43' 4" | 31' 5" | 11' 7" | 1917–18 |
| .. | .. | .. | .. | .. | 4 | .. | 1,800 | 1,100 | 250 | 450 | 468 | 36' | 29' 6" | 10' 5" | 1914–15 |
| .. | .. | .. | .. | .. | 3 | 13,000 | 1,830 | 1,100 | 280 | 450 | 335 | 36' | 29' 6" | 11' 0" | 1917–18 |
| .. | .. | .. | .. | .. | .. | .. | .. | .. | .. | .. | .. | .. | .. | .. | 1914–15 |
| 108 | .. | .. | .. | .. | 3¼ | 10,000 | 2,142 | 1,370 | 252 | 520 | 371 | 37' | 27' 9" | 11' 4" | 1915–17 |
| 55 | .. | .. | .. | .. | 3¼ | 11,000 | 2,100 | 1,431 | 239 | 430 | 360 | 40' 9" | 27' 3" | 12' 0" | 1916–17 |
| .. | .. | .. | .. | .. | .. | .. | .. | .. | .. | .. | .. | .. | .. | .. | 1914–15 |
| 260 | .. | .. | .. | .. | 3 | 12,500 | 2,104 | 1,540 | 283 | 281 | 371 | 37' | 27' 3" | 11' 4" | 1916–17 |
| .. | .. | .. | .. | .. | .. | .. | .. | .. | .. | .. | .. | .. | .. | .. | 1918 |
| 115 | .. | .. | .. | .. | .. | 10,500 | 8,017 | 5,284 | 1,730 | 1,003 | 868 | 74' 10" | 42' 6" | 16' 0" | 1918 |
| .. | .. | .. | .. | .. | .. | .. | 1,388 | 770 | 200 | 418 | 248 | 34' 3" | .. | .. | 1914 |
| 300 | 50 | 115 | .. | .. | 2½ | 15,500 | 1,195 | 757 | 178 | 260 | 198 | 24' 5" | 20' 7" | 8' 6" | 1914–16 |
| 645 | 21·3 | 375 | .. | .. | 3 | 20,000 | 2,779 | 1,934 | 300 | 545 | 406 | 39' 4" | 25' 10" | 9' 9" | 1917–18 |
| 475 | 28·8 | 270 | .. | .. | .. | 19,000 | 2,630 | 1,733 | 345 | 552 | 405 | 39' 3" | 24' 9" | 9' 6" | 1918 |
| 670 | 19·8 | 395 | .. | .. | 1¾ | 20,000 | 1,348 | 896 | 192 | 260 | 140 | 30' 10" | 20' 6" | 7' 10" | 1917–18 |
| .. | .. | .. | .. | .. | 4 | 10,000 | 1,619 | 981 | 253 | 385 | 304 | 43' 5" | 22' 6" | 8' 5" | 1914–15 |
| .. | .. | .. | .. | .. | 3¼–4 | 14,000 | 2,970 | 1,870 | 383 | 717 | 427·5 | 56' 5" | 23' 6" | 8' 5" | 1915–16 |
| .. | .. | .. | .. | .. | 4 | .. | 2,130 | 1,580 | 168 | 382 | 346 | 43' 9" | 27' 1" | 10' 10" | 1915–17 |
| 220 | .. | .. | .. | .. | .. | 13,500 | 2,340 | 1,610 | 330 | 400 | 405 | 41' 4" | 29' 3" | 10' 3" | 1915–17 |

| Aircraft | Type | No. | Engine | Speed (at altitude) | | | | | Climb | Ceiling | Endurance |
|---|---|---|---|---|---|---|---|---|---|---|---|
| de Havilland 2 (D.H.2: Pusher-biplane) | Fighter | 1 | 100-h.p. Monosoupape | 105 (at 1,200) | 86 | 77 | .. | .. | 12 | 380 | 24.8 |
| de Havilland 4 (D.H.4: Tractor-biplane) | Day Bomber and Fighter-Reconnaissance | 2 | 200-h.p. R.A.F. 3a | 235 (at 1,800) | 120 | 117 | 110.5 | .. | 8.0 | 650 | 14.2 |
| | | | 250-h.p. Rolls-Royce | 285 (at 1,800) | 117 | 113 | 102.5 | .. | 8.9 | 550 | 16.4 |
| | | | 375-h.p. Rolls-Royce | 375 (at 2,000) | 136.5 | 133.5 | 126 | 122.5 | 5.2 | 1,042 | 9.0 |
| de Havilland 5 (D.H.5: Tractor-biplane) | Fighter | 1 | 110-h.p. Le Rhône | 126 (at 1,250) | .. | 102 | 89 | .. | 6.9 | 745 | 12.4 |
| de Havilland 6 (D.H.6: Tractor-biplane) | Training and Anti-Submarine patrol | 2 | 90-h.p. R.A.F. 1a | 100 (at 1,800) | 66 | .. | .. | .. | 29 | .. | .. |
| de Havilland 9 (D.H.9: Tractor-biplane) | Day Bomber | 2 | 200-h.p. B.H.P. | 240 (at 1,400) | .. | 111.5 | 97.5 | 91 | 10 | 500 | 20 |
| | | | | (Performance with bomb load) | | | | | | | |
| de Havilland 9a (D.H.9a: Tractor-biplane) | Day Bomber | 2 | 350 Rolls-Royce (Eagle 8) | 359 (at 1,800) | .. | 107 | .. | .. | 14.7 | 345 | 27 |
| | | | | (Performance with bomb load) | | | | | | | |
| | | | 400-h.p. Liberty | 405 (at 1,650) | .. | 114 | 106 | 102 | 8.9 | 595 | 15.8 |
| | | | | (Performance with bomb load) | | | | | | | |
| de Havilland 10 (D.H.10: Tractor-biplane) | Bomber | 3 | 2–400-h.p. Liberty | 2—405 (at 1,600) | 117 | 115 | 110 | .. | 8.2 | 650 | 14.0 |
| F.E.2b (Pusher-biplane) | Fighter-Reconnaissance and Night Bomber | 2 | 120-h.p. Beardmore | 133 (at 1,200) | 73 | 72 | .. | .. | 19.5 | 240 | 45.2 |
| | | | | (Performance without bombs) | | | | | | | |
| | | | 160-h.p. Beardmore | 170 (at 1,350) | 81 | 76 | .. | .. | 18.9 | 210 | 39.7 |
| | | | | (Performance without bombs) | | | | | | | |
| F.E.2d (Pusher-biplane) | Fighter-Reconnaissance | 2 | 250-h.p. Rolls-Royce | 250 (at 1,600) | 93 | 88 | .. | .. | 10.0 | 500 | 18.0 |
| F.E.8 (Pusher-biplane) | Fighter | 1 | 100-h.p. Monosoupape | 105 (at 1,200) | 94.5 (ground level) | .. | .. | .. | 11 | 400 | 17.5 |
| Farman, Henri (Pusher-biplane) | General Purpose | 2 | 80-h.p. Gnome | 84 (at 1,200) | 60 (ground level) | .. | .. | .. | 18.5 (to 3,000) | .. | .. |
| Farman, Henri 'All-steel' | General Purpose | 2 | 140-h.p. Salmson (Canton-Unné) | 140 (at 1,350) | 90 | 88 | .. | .. | 14 | .. | 25 |
| Farman, Maurice (Pusher-biplane) | Training | 2 | 70-h.p. Renault | 75 (at 1,200) | 66 (ground level) | .. | .. | .. | 15 (to 3,000) | .. | .. |
| Handley Page (Tractor-biplane) 0·400 | Night Bomber | 4 | 2–275-h.p. Rolls-Royce | 2—322 (at 1,800) | 79.5 | .. | .. | .. | 30 | 120 | .. |
| | | | | (Performance with bomb load—16 112-lb. bombs) | | | | | | | |
| Handley Page V.1,500 (Pusher-Tractor biplane) | Night Bomber | 6 | 4 Rolls-Royce Eagle VIII | 4—350 (at 1,800) | 99 | 95 | .. | .. | 10 | .. | 21 |
| Martinsyde (Tractor-biplane) | Fighter and Day Bomber | 1 | 120-h.p. Beardmore | 133 (at 1,200) | 95 | 87 | .. | .. | 10 | 400 | 19 |
| | | | 160-h.p. Beardmore | 170 (at 1,350) | 102 | 99.5 | 94 | .. | 8.6 | 575 | 15.2 |
| Martinsyde (Tractor-biplane) F.4 | Fighter | 1 | 300-h.p. Hispano-Suiza | 305 (at 1,800) | 144.5 | 142.5 | 136.5 | .. | 4.0 | 1,415 | 6.2 |
| Morane Scout (Tractor-monoplane) | Fighter | 1 | 80 and 110-h.p. Le Rhône | 84–113 (at 1,200) | 102.4 | 96 | .. | .. | 6.45 | .. | 12 |
| | | | | (Performance with higher horse-power engine) | | | | | | | |

| | | | | | | | | | | | | | | | | |
|---|---|---|---|---|---|---|---|---|---|---|---|---|---|---|---|---|
| | 470 | 29·3 | 220 | 38 | 170 | 4 | 19,500 | 3,340 | 2,304 | 510 | 526 | 436 | 42' 6" | 29' 8" | 10' 5" | 1917–18 |
| | 380 | 36·7 | 150 | 46 | 125 | 3½ | 18,000 | 3,313 | 2,303 | 465 | 545 | 436 | 42' 6" | 30' 8" | 10' 5" | |
| | 830 | 16·5 | 525 | 20 | 450 | 3¾ | 23,000 | 3,472 | 2,403 | 524 | 545 | 436 | 42' 6" | 29' 8" | 11' | |
| | 540 | 27·5 | 165 | .. | .. | 2¾ | 16,000 | 1,492 | 1,010 | 222 | 260 | 211 | 25' 8" | 22' 4" | 8' 5" | 1917–18 |
| | .. | .. | .. | .. | .. | .. | .. | 2,027 | 1,460 | 207 | 360 | 436 | 35' 11" | 27' 3½" | 10' 9¾" | 1917–18 |
| | 343 | 45 | 135 | 67 | 58 | 4½ | 17,500 | 3,669 | 2,203 | 535 | 931 (including 2 230 lb. bombs) | 436 | 42' 6" | 30' 6" | 10' | 1918 |
| | 235 | .. | .. | .. | .. | 6 | 13,500 | 5,000 | 2,732 | 1,060 | 1,208 (including 660 lb. weight bombs) | 488 | 46' | 30' | 10' 9" | 1918 |
| | 425 | 33 | 185 | 43·8 | 110 | 5¾ | 16,500 | 4,645 | 2,695 | 910 | 940 (including 2 230 lb. bombs) | 493 | 46' 0" | 30' | 11' 4" | 1918 |
| | 470 | 29·9 | 220 | .. | .. | 6 | 17,500 | 8,500 | 5,585 | 1,710 | 1,205 | 851 | 64' 9" | 39' 9" | 13' 6" | 1918 |
| | 50 | .. | .. | .. | .. | 3½ | 9,000 | 2,827 | 2,105 | 270 | 452 | 494 | 47' 10" | 32' 3" | 12' 8" | 1915–17 |
| | 140 | .. | .. | .. | .. | | 11,000 | 3,037 | 2,121 | 396 | 520 | 494 | 47' 10" | 32' 3" | 12' 8" | 1917–18 |
| | 302 | .. | .. | .. | .. | 3 | 16,500 | 3,469 | 2,509 | 520 | 440 | 494 | 47' 10" | 32' 3" | 12' 8" | 1916–17 |
| | 225 | .. | .. | .. | .. | | .. | 1,346 | 895 | 232 | 219 | 218 | 31' 6" | 23' 8" | 9' 2" | 1915–17 |
| | .. | .. | .. | .. | .. | .. | .. | .. | .. | .. | .. | .. | .. | .. | .. | 1914 |
| | .. | .. | .. | .. | .. | 4 | .. | .. | .. | .. | .. | .. | 53' | 30' 3" | 12' | 1915–16 |
| | .. | .. | .. | .. | .. | 3¾ | .. | 2,046 | 1,441 | 242 | 363 | 561 | 53' | 30' 8" | 10' 4" | 1914–17 |
| | .. | .. | .. | .. | .. | 8 | 7,000 | 14,022 | 8,480 | 2,830 | 2,712 (including 16 112 lb. bombs) | 1,642 | 100' | 63' | 22' | 1917–18 |
| | .. | .. | .. | .. | .. | 14 | .. | 30,000 | 15,000 | 7,000 | 8,000 (including 30 250 lb. bombs) | 3,000 | 126' | 62' | 23' | 1918 |
| | 250 | .. | .. | .. | .. | 5½ | 14,000 | 2,424 | 1,759 | 421 | 244 | 410 | 38' 1" | 27' | 9' 8" | 1916–17 |
| 2 | 450 | .. | .. | .. | .. | 4½ | 16,000 | 2,458 | 1,793 | 389 | 276 | 410 | 38' 1" | 27' | 9' 8" | |
| 7 | 1,175 | 11·8 | 830 | .. | .. | 2½ | 26,000 | 2,289 | 1,710 | 298 | 281 | 329·5 | 32' 9" | 25' 3" | 8' | 1918 |
| | .. | .. | .. | .. | .. | 1¾ | 13,000 | 1,122 | 735 | 116 | 271 | 118 | 27' 5" | 22' 7" | 8' 3" | 1914–16 |
| 30 | .. | .. | .. | .. | .. | 2½ | 15,000 | 1,612 | 952 | 209 | 451 | 193·7 | 36' 9" | 23' 7" | 11' 5" | 1916–17 |

| Aircraft | Role | No. | Engine | Speed | | | | | | | | |
|---|---|---|---|---|---|---|---|---|---|---|---|---|
| Morane 'Parasol' (Tractor-monoplane) | Corps-Reconnaissance | 2 | 80- and 110-h.p. Le Rhône | 84–113 (at 1,200) | 96 | .. | .. | .. | 8.45 | .. | 15.30 | .. |
| | | | | | Performance with higher horse-power engine) | | | | | | | |
| Nieuport Scout (Tractor-biplane) | Fighter | 1 | 110-h.p. Le Rhône | 113 (at 1,200) | 107 | 101 | .. | .. | 5.5 | .. | 9 | .. |
| Nieuport (Tractor-biplane) | Corps-Reconnaissance | 2 | 110-h.p. Clerget | 113 (at 1,200) | 91 | .. | .. | .. | 14.25 | .. | .. | .. |
| R.E.5 (Tractor-biplane) | Day Bomber and Reconnaissance | 2 | 120-h.p. Beardmore | 133 (at 1,200) | 78 (ground level) | .. | .. | .. | 14.7 (to 6,000) | .. | .. | .. |
| R.E.7 (Tractor-biplane) | Day Bomber | 2 | 150-h.p. R.A.F. 4a | 160 (at 1,800) | 82 | .. | .. | .. | 37.8 | 115 | .. | .. |
| | | | 160-h.p. Beardmore | 170 (at 1,350) | 91 (ground level) | .. | .. | .. | 18.5 | 288 | 31.8 | 279 |
| | | | 250-h.p. Rolls-Royce | 278 (at 1,800) | .. | 88 | .. | .. | 13 | 500 | 23.5 | 425 |
| R.E.8 (Tractor-biplane) | Corps-Reconnaissance | 2 | 150-h.p. R.A.F. 4a | 160 (at 1,800) | 102 | 96 | .. | .. | 15 | 340 | 29 | 235 |
| S.E.5 (Tractor-biplane) | Fighter | 1 | 150-h.p. Hispano-Suiza | 150 (at 1,400) | 119 | 114 | 98 | .. | 8 | 650 | 14.2 | 480 |
| S.E.5a (Tractor-biplane) | Fighter | 1 | 200-h.p. Wolseley Viper | 210 (at 2,000) | 132 | 128 | 115.5 | 107.5 | 6 | 765 | 11.3 | 580 |
| Short Bomber (Tractor-biplane) | Bomber | 2 | 250-h.p. Rolls-Royce | 278 (at 1,800) | 77.5 | .. | .. | .. | 21.4 | .. | 45 | .. |
| Sopwith 'Tabloid' (Tractor-biplane) | Fighter | 1 | 80-h.p. Gnome | 84 (at 1,200) | 92 (ground level) | .. | .. | .. | .. | .. | .. | .. |
| Sopwith 1½ strutter (Tractor-biplane) | Fighter-Reconnaissance | 2 | 110- and 130-h.p. Clerget | 122–125 (at 1,250) | 100 | 97.5 | 87 | .. | 9.2 | 540 | 17.8 | 316 |
| | Day Bomber | | 110- and 130-h.p. Clerget | 122–125 (at 1,250) | 102 | 98.5 | .. | .. | 12.7 | 380 | 24.6 | 222 |
| Sopwith 'Pup' (Tractor-biplane) | Fighter | 1 | 80-h.p. Le Rhône | 84 (at 1,200) | 106.5 | 104.5 | 94 | .. | 8.0 | 650 | 14.4 | 455 |
| Sopwith Triplane (Tractor-triplane) | Fighter | 1 | 130-h.p. Clerget | 125 (at 1,250) | 116 | 114 | 105 | .. | 6.3 | 810 | 10.5 | 790 |
| Sopwith 'Camel' (Tractor-biplane) | Fighter | 1 | 110-h.p. Le Rhône | 137 (at 1,250) | .. | 118.5 | 111.5 | .. | 5.2 | 1,000 | 9.2 | 780 |
| | | | 130-h.p. Clerget | 125 (at 1,250) | .. | 113 | 106.5 | .. | 6.0 | 880 | 10.6 | 665 |
| | | | 150-h.p. B.R.1 | 150 (at 1,250) | .. | 115 | 110 | .. | 5.5 | 995 | 9.4 | 770 |
| | | | 150-h.p. Mono. Gnome | 150 (at 1,225) | .. | 117.5 | 107 | 104.5 | 5.8 | 890 | 10.3 | 675 |
| Sopwith 'Dolphin' (Tractor-biplane) | Fighter | 1 | 200-h.p. Hispano-Suiza | 210 (at 1,800) | .. | 128 | 119.5 | .. | 6.4 | 855 | 11 | 675 |
| Sopwith 'Snipe' (Tractor-biplane) | Fighter | 1 | 200-h.p. B.R.2 | 228 (at 1,300) | .. | 121 | 113 | 108.5 | 5.2 | 970 | 9.4 | 710 |
| Sopwith 'Salamander' (Tractor-biplane) | Armoured Ground Fighter | 1 | 200-h.p. B.R.2 | 228 (at 1,300) | 123.5 | 117 | .. | .. | 9.1 | 550 | 17.1 | 335 |
| S.P.A.D. (Tractor-biplane) | Fighter | 1 | 150-h.p. Hispano-Suiza | 150 (at 1,400) | 119 | 115.5 | 107.5 | .. | 6.3 | 810 | 11.5 | 570 |
| Vickers Scout (Tractor-biplane) | Fighter | 1 | 110-h.p. Le Rhône | 113 (at 1,250) | 109 | at 8,000' | .. | .. | 7.7 | 605 | 14.8 | 465 |
| Vickers Fighter (Pusher-biplane) | Fighter-Reconnaissance | 2 | 100-h.p. Monosoupape | 105 (at 1,200) | 79 | 75 | .. | .. | 19 | .. | 51 | .. |
| Voisin (Pusher-biplane) | Bomber | 2 | 140-h.p. Salmson (Canton-Unné) | 140 (at 1,350) | 62 | .. | .. | .. | 24.5 | .. | .. | .. |

| | | | | | | | | | | | | | | | | |
|---|---|---|---|---|---|---|---|---|---|---|---|---|---|---|---|---|
| 9 | .. | .. | .. | .. | .. | 2 | 17,400 | 1,232 | 825 | 143 | 264 | 158·8 | 26' | 19' | 7' | 1916–18 |
| .. | .. | .. | .. | .. | .. | 3 | 13,000 | 1,870 | 1,210 | 264 | 396 | 236·5 | 29' 7" | 23' | 8' 9" | 1916–17 |
| .. | .. | .. | .. | .. | .. | .. | .. | .. | .. | .. | .. | .. | .. | .. | .. | 1914–15 |
| .. | .. | .. | .. | .. | .. | 6 | 6,500 | 3,449 | 2,170 | 549 | 730 | | | | | |
| 31·8 | 279 | .. | .. | .. | .. | .. | .. | 3,290 | 2,285 | 485 | 520 | 548 | 57' | 31' 10" | 12' 7" | 1915–17 |
| 23·5 | 425 | .. | .. | .. | .. | .. | .. | 4,109 | 2,702 | 605 | 802 | | | | | |
| 29 | 235 | .. | .. | .. | .. | 4¼ | 13,000 | 2,678 | 1,803 | 355 | 520 | 377·5 | 42' 7" | 27' 10" | 11' 4" | 1916–18 |
| 14·2 | 480 | 29·5 | 210 | .. | .. | 2½ | 17,000 | 1,930 | 1,399 | 245 | 286 | 249 | 28' | 21' 4" | 9' 5" | 1917–18 |
| 11·3 | 580 | 22·9 | 305 | .. | .. | 2½ | 20,000 | 2,048 | 1,531 | 230 | 287 | 247 | 26' 8" | 21' | 9' 6" | 1917–18 |
| 45 | .. | .. | .. | .. | .. | 6 | 9,500 | .. | .. | .. | .. | 870 | 85' | 36' | 15' | 1916–17 |
| .. | .. | .. | .. | .. | .. | .. | .. | .. | .. | .. | .. | .. | .. | .. | .. | 1914 |
| 17·8 | 316 | 41·9 | 135 | .. | .. | 3¾ | 15,500 | 2,150 | 1,305 | 328 | 517 | 346 | 33' 6" | 25' 3" | 10' 3" | 1916–17 |
| 24·6 | 222 | .. | .. | .. | .. | .. | 13,000 | 2,342 | 1,316 | 502 | 524 | | | | | |
| 14·4 | 455 | 30·1 | 215 | .. | .. | 3 | 17,500 | 1,225 | 787 | 178 | 260 | 254 | 26' 9" | 19' 7" | 9' | 1916–17 |
| 10·5 | 790 | 19·0 | 460 | .. | .. | 2¾ | 20,500 | 1,415 | 993 | 184 | 238 | 257 | 26' 7" | 19' 6" | 9' 9" | 1916–17 |
| 9·2 | 780 | 16·8 | 540 | .. | .. | 2¾ | 24,000 | 1,422 | 889 | 252 | 281 | 231 | 28' | 18' 8" | 8' 6" | |
| 10·6 | 665 | 20·7 | 355 | .. | .. | 2½ | 19,000 | 1,453 | 929 | 243 | 281 | 231 | 28' | 18' 8" | 8' 6" | |
| 9·4 | 770 | 18·0 | 440 | .. | .. | .. | 20,000 | 1,471 | 962 | 228 | 281 | 231 | 28' | 18' 8" | 8' 6" | 1917–18 |
| 10·3 | 675 | 19·7 | 410 | 23·7 | 335 | 2¼ | 21,000 | 1,523 | 993 | 249 | 281 | 233 | 28' 2" | 19' | 8' 6" | |
| 11 | 675 | 20·2 | 420 | 23·9 | 345 | 2¼ | 21,000 | 2,000 | 1,391 | 328 | 281 | 262 | 32' 6" | 22' 6" | 7' 9" | 1918 |
| 9·4 | 710 | 18·8 | 390 | 23·2 | 290 | 3 | 20,000 | 2,020 | 1,312 | 343 | 365 | 270 | 30' 1" | 19' 9" | 8' 9" | 1918 |
| 17·1 | 335 | .. | .. | .. | .. | .. | 13,000 | 2,512 | 1,844 | 258 | 410 | 267 | 30' | 18' 9" | 8' 7" | 1918 |
| 11·5 | 570 | 24·5 | 240 | .. | .. | 2½ | 17,500 | 1,632 | 1,177 | 195 | 260 | 200 | 25' 6" | 20' 3" | 7' | 1917–18 |
| 14·8 | 465 | 32·5 | 256 | .. | .. | 3¼ | 10,000 | 1,617 | 1,068 | 300 | 249 | 215 | 24' 6" | 19' 10" | 7' 8" | 1916 |
| 51 | .. | .. | .. | .. | .. | 5 | 11,000 | 1,892 | 1,029 | 440 | 423 | .. | 33' 10" | 27' 10" | 7' | 1915–16 |
| .. | .. | .. | .. | .. | .. | 4 | 10,000 | 2,959 | 2,090 | 358 | 511 | 535 | 48' 5" | 31' 3" | 9' 8" | 1915–16 |

# APPENDIX XXVII

## TABLE B. SEAPLANES AND SHIP AEROPLANES

| Type. | Tractor or Pusher. | No. of Seats. | Engine. | Normal B.H.P. and R.P.M. at Ground Level. | Lifting Surface. | Speed in Knot At 2,000 ft. | At 6,500 ft. |
|---|---|---|---|---|---|---|---|
| *Ship Aeroplanes.* | | | | | | | |
| Parnall 'Panther' | T | 2 | 230 B.R.2 | 228 (at 1,300) | 325 | .. | 94 |
| Sopwith 'Camel' | T | 1 | 150 B.R.1 | 150 (at 1,250) | 229 | 108 | 105·5 |
| Sopwith 'Torpedo' | T | 1 | Sunbeam 'Arab' | 207 (at 2,000) | 568 | 90·5 | 89 |
| *Float Seaplanes.* | | | | | | | |
| Fairey 'Campania' | T | 2 | 275 Rolls-Royce Mk. I. | 307 (at 1,800) | 654 | 76 | 72 |
| ,,    ,, | T | 2 | 260 Sunbeam | 265 (at 2,100) | 655 | 73·5 | 68 |
| Fairey 3b | T | 2 | Sunbeam 'Maori' | 265 (at 2,100) | 570 | 79 | 75·5 |
| Fairey 3c (Light Load) | T | 2 | Rolls-Royce 'Eagle' 8 | 359 (at 1,800) | 476 | 96 | 93 |
| Fairey 3c (Medium Load) | T | 2 | ,,    ,, | 359 (at 1,800) | 476 | .. | 91 |
| Fairey 3c (Normal Load) | T | 2 | ,,    ,, | 359 (at 1,800) | 476 | 96 | 93 |
| Fairey 3c (Over Load) | T | 2 | ,,    ,, | 359 (at 1,800) | 476 | 87·5 | 83 |
| Short Improved 184 | T | 2 | 240 Renault | 225 (at 1,200) | 670 | 69·5 | 61 |
| ,,    ,, | T | 2 | 260 Sunbeam | 265 (at 2,100) | 680 | 73 | 72 |
| *Boat Seaplanes.* | | | | | | | |
| F.2a | T | 4 | 2 Rolls-Royce 'Eagle' 8 | 2—345 (at 1,800) | 1,133 | 83 | 77 |
| F.3 (Light Load) | T | 4 | ,,    ,, | ,,    ,, | 1,430 | 81 | 79·5 |
| F.3 (Medium Load) | T | 4 | ,,    ,, | ,,    ,, | 1,430 | 80·5 | .. |
| F.3 (Normal Load) | T | 4 | ,,    ,, | ,,    ,, | 1,430 | 79 | 74·5 |
| F.3 (Over-Load) | T | 4 | ,,    ,, | ,,    ,, | 1,430 | 78 | 75·5 |
| F.5 (Light Load) | T | 4 | ,,    ,, | ,,    ,, | 1,409 | 88 | 87 |
| F.5 (Medium Load) | T | 4 | ,,    ,, | ,,    ,, | 1,409 | 86 | 85 |
| F.5 (Normal Load) | T | 4 | ,,    ,, | ,,    ,, | 1,409 | 89 | 86 |
| F.5 (Over-Load) | T | 4 | ,,    ,, | ,,    ,, | 1,409 | 88·5 | 80·5 |
| H.16 | T | 4 | ,,    ,, | ,,    ,, | 1,200 | 85·5 | 83·5 |
| N.T.2b (School) | P | 2 | Sunbeam 'Arab' | 212 (at 2,000) | 453 | 74 | 73 |

\* 18 in. Torpedo.

NOTE.—Lifting Surface = Surface of Wings and Flaps only.
Military Load = Weight of Guns, Bombs, Ammunition, and Reconnaissance Load.
Air Endurance = At 10,000 ft. Alt., at full throttle, including climb.
Service Ceiling = Height at which Rate of Climb is 100 ft. per min.
Weight Empty = Includes Cooling Water for Water-cooled engines.

| Time in Mins. and Rate of Climb in Feet per Min. | | | | | | Air Endurance. | Service Ceiling. Ft. | Weight. | | | | |
|---|---|---|---|---|---|---|---|---|---|---|---|---|
| 2,000 ft. | | 6,500 ft. | | 10,000 ft. | | | | Gross. | = Empty. | + Fuel and Oil. | + Military Load. | + Crew. |
| ime. | Rate. | Time. | Rate. | Time. | Rate. | | | | | | | |
| ·3 | 795 | 9·3 | 545 | 17·1 | 345 | 4½ (at 10,000 ft.) | 14,500 | 2,595 | 1,328 | 541 | 366 | 360 |
| ·8 | 1,278 | 6·2 | 752 | 11·4 | 541 | .. | 17,500 | 1,530 | 1,036 | 223 | 91 | 180 |
| | 466 | 15·7 | 303 | 31 | 176 | 4 (at 10,000 ft.) | 12,000 | 3,883 | 2,199 | 405 | 1,099* | 180 |
| ·6 | 307 | 28·5 | 120 | .. | .. | 4½ | 7,000 | 5,406 | 3,613 | 783 | 650 | 360 |
| | 242 | 38 | 80 | .. | .. | 4½ | 6,000 | 5,329 | 3,672 | 631 | 666 | 360 |
| ·3 | 410 | 19·5 | 210 | .. | .. | | 9,000 | 5,083 | 3,423 | 610 | 690 | 360 |
| ·3 | 910 | 7·9 | 650 | 14·3 | 460 | 2 (at 6,000 ft.) | 17,000 | 4,272 | 3,392 | 350 | 170 | 360 |
| ·7 | 848 | 8·9 | 545 | 17·5 | 369 | 4 (at 6,000 ft.) | 15,500 | 4,600 | 3,392 | 678 | 170 | 360 |
| ·3 | 794 | 9·3 | 510 | 18 | 332 | 5½ (at 6,000 ft.) | 15,000 | 4,800 | 3,392 | 878 | 170 | 360 |
| ·7 | 500 | 16·5 | 246 | 44 | 48 | 5 (at 6,000 ft.) | 8,500 | 5,039 | 3,549 | 883 | 247 | 360 |
| ·7 | 215 | 42·5 | 75 | .. | .. | 4½ | 5,500 | 5,560 | 3,798 | 734 | 668 | 360 |
| ·3 | 300 | 26·3 | 165 | .. | .. | 4½ | 9,000 | 5,123 | 3,479 | 637 | 647 | 360 |
| ·8 | 470 | 16·7 | 252 | 39·5 | 86 | 6 (at 1,000 ft.) | 9,500 | 10,978 | 7,549 | 2,124 | 585 | 720 |
| ·2 | 580 | 12·9 | 378 | 24·8 | 218 | 2¼ (at 2,000 ft.) | 12,500 | 9,752 | 7,958 | 836 | 238 | 720 |
| | 440 | 18 | 240 | 41·5 | 90 | 6 (at 2,000 ft.) | 10,000 | 11,084 | 7,958 | 2,089 | 317 | 720 |
| ·4 | 333 | 24 | 163 | .. | .. | 6 (at 2,000 ft.) | 8,000 | 12,235 | 7,958 | 2,096 | 1,461† | 720 |
| ·8 | 230 | 41 | 75 | .. | .. | .. | 6,000 | 13,281 | 7,958 | 3,142 | 1,461† | 720 |
| ·8 | 701 | 10·2 | 525 | 17·8 | 389 | 2¼ (at 6,000 ft.) | 17,500 | 9,630 | 8,023 | 688 | 199 | 720 |
| ·7 | 525 | 14 | 355 | 26·2 | 228 | 6·87 (at 6,500 ft.) | 13,500 | 11,337 | 8,023 | 2,086 | 508 | 720 |
| | 462 | 16·1 | 290 | 32·5 | 160 | 7 (at 6,000 ft.) | 11,500 | 12,268 | 8,023 | 2,097 | 1,428† | 720 |
| ·3 | 352 | 22·5 | 193 | .. | .. | 7 (at 6,000 ft.) | 9,000 | 13,306 | 8,023 | 3,121 | 1,442† | 720 |
| ·7 | 512 | 14·6 | 335 | 28 | 198 | 6 (at 2,000 ft.) | 12,500 | 10,670 | 7,363 | 2,115 | 472 | 720 |
| ·1 | 445 | 16·7 | 280 | 33·7 | 150 | .. | 11,500 | 3,169 | 2,321 | 488 | .. | 360 |

† With 4—230 lb. bombs.

## APPENDIX XL

### STRENGTH OF ALLIED AIRCRAFT ON ALL FRONTS: JUNE 1918

| Type. | Western Front. ||||||| Italian Front. |||| Macedonian Front. ||||| Middle East. |||| Mediterranean. ||| Home Defence. ||| Grand Totals. |
|---|---|---|---|---|---|---|---|---|---|---|---|---|---|---|---|---|---|---|---|---|---|---|---|---|---|---|---|
| | British ||| French. | Belgian. | American. | Total. | Italian. | British. | Total. | British. | French. | Italian. | Total. | British. | Italian (Front d'Orient). | Total. | British. | Total. | British. | French. | Total. | |
| | 5th Group Belgian Coast. | Indepent Force. | | | | | | | | | | | | | | | | | | | | |
| *Corps.* | | | | | | | | | | | | | | | | | | | | | | | |
| Reconnaissance | 480 | .. | 1,555 | .. | 72 | 2,107 | 201 | 24 | 225 | 26 | 130 | 13 | 169 | 190 | 9 | 199 | .. | .. | .. | 148 | 148 | 2,848 |
| *Fighter.* | | | | | | | | | | | | | | | | | | | | | | | |
| Single-seater Fighter | 720 | 48 | 1,050 | 69 | 90 | 1,977 | 223 | 72 | 295 | 15 | 205 | 11 | 231 | 49 | 15 | 64 | 72 | 72 | 144 | 105 | 249 | 2,888 |
| Two-seater Reconnaissance | 144 | .. | 103 | 58 | .. | 305 | .. | 8 | 8 | .. | .. | .. | .. | 24 | .. | 24 | .. | .. | 48 | .. | 48 | 385 |
| *Bomber.* | | | | | | | | | | | | | | | | | | | | | | | |
| Day | 144 | 36 | 218 | .. | 18 | 452 | 45 | .. | 45 | .. | 18 | .. | 18 | 6 | 4 | 10 | 72 | 72 | .. | 1 | 1 | 598 |
| Short Night Bomber | 118 | .. | 161 | .. | .. | 279 | .. | .. | .. | .. | 10 | .. | 10 | .. | .. | .. | .. | .. | .. | 91 | 91 | 380 |
| Long Night Bomber | .. | 10 | 16 | .. | .. | 26 | 8 | .. | 8 | .. | .. | .. | .. | .. | .. | .. | .. | .. | .. | .. | .. | 34 |
| Medium Night Bomber | .. | .. | 46 | .. | .. | 46 | .. | .. | .. | .. | .. | .. | .. | .. | .. | .. | .. | .. | .. | .. | .. | 46 |
| *Night Flying.* | | | | | | | | | | | | | | | | | | | | | | | |
| | 1,606 | 84 | | | | | | | | | | | | | | | | | | | | |
| | | 46 | | | | | | | | | | | | | | | | | | | | |
| Total | 1,736 || 3,149 | 127 | 180 | 5,192 | 477 | 104 | 581 | 41 | 363 | 24 | 428 | 269 | 28 | 297 | 144 | 144 | 336 | 345 | 681 | 7,323 |

British . . . . 2,630
French . . . . 3,857
American . . . . 180
Italian . . . . 529
Belgian . . . . 127

# APPENDIX XLI

DISPOSITION OF AIRCRAFT AND ENGINES ON CHARGE OF THE ROYAL AIR FORCE AT 31st OCTOBER 1918

## TABLE A. AEROPLANE AND SEAPLANES (AIRFRAMES)

| Type. | In transit to and at Aeroplane Repair Depots. | In Store. | At Aircraft Acceptance Parks and Contractors. | Squadrons Mobilising. | At Home. Areas. | Schools. | 6th Brigade. | 11th (Irish) Group. | Sundry Units. | C.T.D. |
|---|---|---|---|---|---|---|---|---|---|---|
| *Aeroplanes.* | | | | | | | | | | |
| Avro | 10 | 16 | 10 | .. | 1,933 | 311 | 226 | 184 | 18 | 5 |
| A.W. (Beardmore) | 21 | 263 | 36 | .. | 18 | 43 | .. | .. | 5 | 3 |
| A.W. (R.A.F.) | 1 | .. | .. | .. | 28 | 23 | .. | .. | .. | 1 |
| B.E. 2 c, d, and e | 3 | 10 | 2 | .. | 150 | 79 | 21 | 4 | 34 | 16 |
| B.E. 12 and 12a | .. | .. | .. | .. | 2 | 3 | 12 | .. | 4 | 4 |
| B.E.12b | 31 | 67 | .. | .. | .. | .. | 17 | .. | .. | .. |
| B.A.T. | .. | .. | 1 | .. | .. | .. | .. | .. | .. | 2 |
| Blackburn 'Kangaroo' | .. | .. | 3 | .. | 10 | 1 | .. | .. | .. | .. |
| Bristol Fighter | 21 | 172 | 46 | 3 | 111 | 62 | 58 | .. | 13 | 11 |
| Bristol Fighter ('Arab') | 62 | 345 | 101 | .. | 95 | 20 | .. | .. | 11 | 8 |
| Bristol Mono | .. | .. | .. | .. | 4 | 26 | .. | .. | 1 | .. |
| Caudron School | .. | .. | .. | .. | .. | 71 | .. | .. | .. | 1 |
| De Havilland 4 (R.R.) | 75 | 23 | 15 | .. | 11 | .. | .. | .. | 8 | 8 |
| De Havilland 4 (R.A.F. 3a) | 6 | .. | 4 | .. | 12 | .. | .. | 3 | 3 | 2 |
| De Havilland 4 (B.H.P.) | 1 | .. | 1 | .. | 35 | 17 | .. | 3 | 5 | 4 |
| De Havilland 4 (Fiat) | .. | .. | .. | .. | 2 | .. | .. | .. | .. | .. |
| De Havilland 6 | 5 | 266 | 2 | .. | 571 | 33 | 33 | 8 | 17 | 6 |
| De Havilland 9 | 24 | 319 | 186 | 37 | 391 | 184 | .. | 41 | 8 | 25 |
| De Havilland 9a | 2 | 24 | 105 | 25 | 17 | 50 | .. | .. | 2 | 19 |
| De Havilland 10 | .. | .. | .. | .. | .. | .. | .. | .. | .. | 6 |
| F.E. 2b and c | 9 | 44 | 68 | .. | 69 | 38 | 97 | .. | 7 | 11 |
| Handley Page 0/400 | 14 | .. | 31 | 4 | 48 | 24 | .. | .. | 5 | 4 |
| Handley Page 'V' Type | .. | .. | .. | .. | .. | .. | .. | .. | .. | 2 |
| Martinsyde F.3 and F.4 | .. | 44 | 1 | .. | .. | .. | .. | .. | .. | 7 |
| R.E.8 | 59 | 282 | 111 | .. | 268 | 167 | 2 | 58 | 4 | 3 |
| S.E.5 and 5a | 35 | 1,407 | 182 | 22 | 268 | 92 | .. | .. | 20 | 20 |
| Sopwith Bomber | .. | 16 | .. | .. | 1 | 6 | .. | .. | 3 | .. |
| Sopwith 'Camel' (B.R.1) | 1 | 84 | .. | .. | 28 | 1 | .. | .. | 1 | 5 |
| Sopwith 'Camel' (Clerget) | 72 | 160 | 25 | .. | 329 | 142 | .. | .. | 13 | 13 |
| Sopwith 'Camel' (Le Rhône and Mono) | 14 | 127 | 119 | .. | 29 | 3 | 181 | .. | .. | 10 |
| Sopwith 'Dolphin' | 38 | 652 | 97 | 8 | 79 | 18 | .. | .. | 7 | 7 |
| Sopwith 'Salamander' | .. | 1 | 31 | 1 | 1 | .. | .. | .. | .. | 1 |
| Sopwith Two-seater | 11 | 18 | .. | .. | 15 | 12 | .. | .. | 2 | 29 |
| Sopwith Scout | 27 | 348 | 1 | .. | 263 | 98 | 34 | .. | 10 | 20 |
| Sopwith 'Snipe' | 2 | .. | 117 | .. | 30 | 5 | 1 | .. | 3 | 14 |
| S.P.A.D. | 3 | 50 | .. | .. | 8 | 3 | .. | .. | 11 | .. |
| Vickers 'Vimy' | .. | .. | .. | .. | .. | .. | .. | .. | .. | 2 |
| Miscellaneous | .. | 1 | .. | .. | .. | 2 | .. | .. | .. | 34 |
| Obsolete | 8 | 135 | 3 | .. | 40 | 26 | .. | .. | 22 | 24 |
| **TOTAL AEROPLANES** | **555** | **4,874** | **1,327** | **100** | **4,865** | **1,560** | **682** | **301** | **237** | **327** |
| *Seaplanes and Ship Aeroplanes.* | | | | | | | | | | |
| 'Campania' (R.R.) | .. | .. | .. | .. | 2 | .. | .. | .. | .. | .. |
| 'Campania' (Sunbeam) | .. | .. | .. | .. | 16 | .. | .. | .. | .. | .. |
| 'Fairey' Type 3a and 3b | .. | 20 | 36 | .. | 10 | .. | .. | .. | .. | 6 |
| F.B.A. Flying-boat | .. | .. | .. | .. | 32 | .. | .. | .. | .. | .. |
| F.2a 'Large America' | .. | .. | .. | .. | 53 | .. | .. | .. | .. | .. |
| F.3 'Large America' | 1 | 17 | 18 | .. | 26 | .. | .. | .. | .. | 3 |
| H.12 'Large America' | .. | .. | .. | .. | 10 | .. | .. | .. | .. | 1 |
| H.12 Converted 'Large America' | 2 | .. | .. | .. | 4 | .. | .. | .. | .. | .. |
| H.16 'Large America' | .. | 31 | 8 | .. | 12 | .. | .. | .. | 13 | 1 |
| Hamble 'Baby' | .. | .. | .. | .. | 9 | .. | .. | .. | .. | .. |
| N.T.2b | .. | 51 | 2 | .. | 23 | .. | .. | .. | .. | 3 |
| Porte Boat | .. | .. | .. | .. | .. | .. | .. | .. | .. | .. |
| S.B.3d | .. | 18 | .. | .. | .. | .. | .. | .. | .. | .. |
| Short 184 (225/240 Sunbeam and Renault) | .. | .. | .. | .. | 13 | .. | .. | .. | .. | .. |
| Short 184 (260 Sunbeam) | .. | .. | 8 | .. | 161 | .. | .. | .. | .. | .. |
| Short 320 | .. | .. | .. | .. | 17 | .. | .. | .. | .. | 2 |
| Short 827 | .. | .. | .. | .. | 3 | .. | .. | .. | .. | .. |
| Sopwith 2F.1 | .. | .. | 1 | .. | 7 | 2 | .. | .. | .. | 1 |
| Sopwith 9901A | .. | .. | .. | .. | .. | .. | .. | .. | .. | .. |
| Sopwith 'Baby' | .. | .. | .. | .. | 34 | .. | .. | .. | .. | .. |
| Sopwith Ship | .. | .. | .. | .. | .. | .. | .. | .. | .. | .. |
| Sopwith Torpedo Plane | .. | .. | 12 | .. | 27 | .. | .. | .. | .. | .. |
| Wight (R.R.) | .. | .. | .. | .. | 2 | .. | .. | .. | .. | .. |
| Wight (Sunbeam) | .. | .. | .. | .. | 5 | .. | .. | .. | .. | .. |
| Miscellaneous | 1 | .. | .. | .. | 3 | .. | .. | .. | .. | 18 |
| Obsolete | .. | 16 | .. | .. | 2 | .. | .. | .. | .. | 4 |
| **TOTAL SEAPLANES AND SHIP AEROPLANES** | **4** | **153** | **85** | **..** | **467** | **2** | **..** | **..** | **13** | **39** |
| **TOTAL ALL MACHINES** | **559** | **5,027** | **1,412** | **100** | **5,332** | **1,562** | **682** | **301** | **250** | **366** |

* Includes all machines not advised as received by Mediterranean and M.E.B. Stations.

| Grand Fleet and Northern Patrol. | Expeditionary Forces. | | | | Awaiting shipment and in transit to, and awaiting assembly in the East. | Eastern Stations. | | | | | Total. | Aircraft written off charge during October. |
|---|---|---|---|---|---|---|---|---|---|---|---|---|
| | E.F. France. | I.F. France. | 14th Wing Italy. | 5th Group. | | Egypt and Palestine. | Salonika. | Mesopotamia. | N.W. Frontier. | Mediterranean (various dates). | | |
| .. | .. | .. | .. | .. | 175 | 111 | .. | .. | .. | .. | 2,999 | 261 |
| .. | 182 | .. | .. | .. | 21 | 56 | 44 | .. | 2 | .. | 694 | 89 |
| .. | .. | .. | .. | .. | .. | 6 | 3 | .. | .. | .. | 62 | 7 |
| 6 | 1 | .. | .. | .. | 7 | 67 | 6 | 6 | 58 | 4 | 474 | 43 |
| .. | 1 | .. | .. | .. | 2 | 26 | 10 | .. | .. | .. | 64 | } 7 |
| .. | .. | .. | .. | .. | .. | .. | .. | .. | .. | .. | 115 | |
| .. | .. | .. | .. | .. | .. | .. | .. | .. | .. | .. | 3 | .. |
| .. | .. | .. | .. | .. | .. | .. | .. | .. | .. | .. | 14 | 1 |
| .. | 249 | .. | 48 | .. | 26 | 42 | .. | .. | .. | .. | 862 | 101 |
| .. | 79 | .. | .. | .. | .. | .. | .. | .. | .. | .. | 721 | 7 |
| .. | .. | .. | .. | .. | .. | 1 | 2 | 13 | .. | .. | 47 | 7 |
| .. | .. | .. | .. | .. | .. | .. | .. | .. | .. | .. | 72 | 3 |
| 4 | 120 | 70 | .. | 37 | .. | .. | .. | .. | .. | 2 | 373 | 48 |
| .. | 25 | .. | .. | .. | 4 | .. | .. | 9 | .. | 4 | 72 | 5 |
| .. | 12 | .. | .. | .. | .. | .. | .. | .. | .. | 20 | 98 | 25 |
| .. | 1 | 2 | .. | .. | .. | .. | .. | .. | .. | .. | 5 | 2 |
| .. | .. | .. | .. | .. | 40 | 69 | .. | .. | .. | .. | 1,050 | 159 |
| 1 | 334 | 71 | .. | 4 | 135 | 7 | 21 | .. | .. | 78 | 1,866 | 192 |
| 12 | 66 | 83 | .. | .. | .. | .. | .. | .. | .. | .. | 405 | 23 |
| .. | .. | 2 | .. | .. | .. | .. | .. | .. | .. | .. | 8 | .. |
| .. | 220 | .. | .. | .. | .. | .. | .. | .. | .. | .. | 563 | 59 |
| .. | 42 | 84 | .. | .. | .. | 2 | .. | .. | .. | .. | 258 | 17 |
| .. | .. | .. | .. | .. | .. | .. | .. | .. | .. | .. | 2 | .. |
| .. | .. | .. | .. | .. | .. | .. | .. | .. | .. | .. | 52 | .. |
| .. | 674 | .. | 49 | .. | 56 | 127 | .. | 53 | .. | .. | 1,913 | 189 |
| .. | 472 | .. | .. | .. | 56 | 91 | 18 | 13 | .. | .. | 2,696 | 223 |
| 2 | .. | .. | .. | 1 | .. | 3 | .. | .. | .. | 2 | 34 | 3 |
| 26 | 186 | .. | .. | 24 | .. | .. | .. | .. | .. | .. | 385 | 64 |
| 4 | 344 | 27 | 75 | 12 | 26 | .. | .. | .. | .. | 100 | 1,342 | 119 |
| 10 | 272 | .. | .. | .. | 25 | 9 | 11 | 11 | .. | .. | 821 | 75 |
| .. | 149 | .. | .. | .. | .. | .. | .. | .. | .. | .. | 1,055 | 55 |
| .. | 2 | .. | .. | .. | .. | .. | .. | .. | .. | .. | 37 | .. |
| 36 | .. | .. | .. | 3 | .. | .. | .. | .. | .. | 11 | 137 | 4 |
| 13 | .. | .. | 1 | .. | 12 | 46 | .. | .. | .. | 5 | 877 | 48 |
| 2 | 88 | 2 | .. | .. | .. | .. | .. | .. | .. | .. | 264 | 9 |
| .. | .. | .. | .. | .. | .. | .. | .. | 15 | .. | .. | 90 | 3 |
| .. | .. | 1 | .. | .. | .. | .. | .. | .. | .. | .. | 3 | .. |
| .. | .. | .. | .. | .. | .. | .. | .. | .. | .. | .. | 37 | 6 |
| .. | 3 | .. | .. | .. | .. | 48 | .. | 3 | .. | 8 | 320 | 95 |
| 116 | 3,522 | 342 | 173 | 81 | 585 | 711 | 115 | 123 | 60 | 234 | 20,890 | 1,949 |
| 24 | .. | .. | .. | .. | .. | .. | .. | .. | .. | .. | 26 | .. |
| .. | .. | .. | .. | .. | .. | .. | .. | .. | .. | .. | 16 | 4 |
| 2 | .. | .. | .. | .. | .. | .. | .. | .. | .. | .. | 70 | 1 |
| .. | .. | .. | .. | .. | .. | .. | .. | .. | .. | .. | 32 | 4 |
| .. | .. | .. | .. | .. | .. | .. | .. | .. | .. | .. | 53 | 1 |
| 18 | .. | .. | .. | .. | .. | .. | .. | .. | .. | 13 | 96 | 2 |
| 1 | .. | .. | .. | .. | .. | .. | .. | .. | .. | .. | 12 | .. |
| .. | .. | .. | .. | .. | .. | .. | .. | .. | .. | .. | 6 | .. |
| 4 | .. | .. | .. | .. | .. | .. | .. | .. | .. | .. | 69 | 1 |
| .. | .. | .. | .. | .. | .. | .. | .. | .. | .. | 9 | 18 | .. |
| .. | .. | .. | .. | .. | .. | .. | .. | .. | .. | .. | 79 | .. |
| 2 | .. | .. | .. | .. | .. | .. | .. | .. | .. | .. | 2 | .. |
| 37 | .. | .. | .. | .. | .. | .. | .. | .. | .. | .. | 55 | 1 |
| 6 | .. | .. | .. | .. | .. | .. | .. | .. | .. | 11 | 30 | 7 |
| 6 | .. | .. | .. | 9 | 35 | .. | .. | .. | .. | 63 | 282 | 18 |
| .. | .. | .. | .. | .. | 1 | .. | .. | .. | .. | .. | 30 | 4 |
| .. | .. | .. | .. | .. | .. | .. | .. | .. | .. | 1 | 4 | .. |
| 112 | .. | .. | .. | .. | .. | .. | .. | .. | .. | 6 | 129 | 15 |
| 10 | .. | .. | .. | .. | .. | .. | .. | .. | .. | .. | 10 | 1 |
| 3 | .. | .. | .. | .. | .. | .. | .. | .. | .. | 21 | 58 | 6 |
| 57 | .. | .. | .. | .. | .. | .. | .. | .. | .. | .. | 57 | 6 |
| 30 | .. | .. | .. | .. | .. | .. | .. | .. | .. | .. | 69 | 6 |
| .. | .. | .. | .. | .. | .. | .. | .. | .. | .. | .. | 2 | 5 |
| .. | .. | .. | .. | .. | .. | .. | .. | .. | .. | .. | 5 | .. |
| 5 | .. | .. | .. | .. | .. | .. | .. | .. | .. | .. | 27 | 4 |
| .. | .. | .. | .. | .. | .. | .. | .. | .. | .. | 2 | 24 | .. |
| 317 | .. | .. | .. | 9 | 36 | .. | .. | .. | .. | 156 | 1,281 | 86 |

TABLE

| Type. | Under and in transit for repair. | In Store. Service. | In Store. Obsolete. | At Home. At Contractors. | At Home. At Aircraft Acceptance Parks. | At Home. Squadrons Mobilising. | At Home. Areas. | At Home. Schools. | At Home. 6th Brigade. | At Home. 11th (Irish) Group. |
|---|---|---|---|---|---|---|---|---|---|---|
| A.B.C. 'Wasp' | .. | .. | .. | 1 | .. | .. | .. | .. | .. | .. |
| A.B.C. 'Dragonfly' | .. | .. | .. | .. | .. | .. | .. | .. | .. | .. |
| 100-h.p. Anzani | .. | .. | 123 | .. | .. | .. | 1 | 117 | .. | .. |
| 120-h.p. Beardmore | 15 | 11 | .. | .. | .. | .. | 32 | 35 | 45 | .. |
| 160-h.p. Beardmore | 319 | 17 | .. | 118 | 107 | .. | 132 | 121 | 77 | 1 |
| 150-h.p. B.R.1 | 81 | .. | .. | 100 | 49 | .. | 47 | 7 | .. | .. |
| 200-h.p. B.R.2 | 22 | 8 | .. | 293 | 155 | 1 | 37 | 9 | 2 | .. |
| 80-h.p. Clerget | 32 | 78 | .. | .. | .. | .. | 73 | 3 | .. | .. |
| 110/130-h.p. Clerget | 297 | 89 | .. | 147 | 45 | .. | 1,438 | 307 | 14 | 162 |
| 140-h.p. Clerget | 75 | 8 | .. | 87 | 28 | .. | 56 | 55 | .. | .. |
| 200-h.p. Clerget | 1 | 1 | .. | .. | .. | .. | .. | .. | .. | .. |
| 90-h.p. Curtiss | 1 | 88 | 218 | 33 | 26 | .. | 276 | 7 | .. | .. |
| 160-h.p. Curtiss | .. | 1 | 34 | .. | .. | .. | 2 | 8 | .. | .. |
| 260-h.p. F.I.A.T. | 42 | 5 | .. | 77 | .. | .. | 3 | .. | .. | .. |
| Galloway Adriatic | 44 | 18 | 9 | .. | 1 | .. | 13 | 2 | .. | .. |
| Galloway Atlantic | 1 | 1 | .. | 16 | .. | .. | 2 | .. | .. | .. |
| 50/80-h.p. Gnome | 4 | 43 | 765 | 9 | 1 | .. | 24 | 4 | .. | .. |
| Hispano-Suiza French | 759 | 48 | .. | 252 | 154 | 9 | 241 | 73 | 13 | .. |
| 300-h.p. Hispano-Suiza | 7 | .. | .. | 6 | 1 | .. | .. | .. | .. | .. |
| 80-h.p. Le Rhône | 166 | 108 | .. | 379 | 4 | .. | 681 | 165 | 47 | 2 |
| 110-h.p. Le Rhône | 380 | 53 | .. | 114 | 53 | .. | 492 | 205 | 468 | .. |
| Liberty 12 | 11 | 17 | .. | 189 | 108 | 28 | 23 | 53 | .. | .. |
| 100 h.p. Monosoupape | 198 | 19 | 4 | 156 | 5 | .. | 642 | 416 | 91 | 24 |
| Napier 'Lion' | .. | 14 | .. | .. | .. | .. | .. | 1 | .. | .. |
| R.A.F.1a and b | 498 | 98 | 14 | 24 | 8 | .. | 862 | 208 | 79 | 16 |
| R.A.F.4a, 4d, and 5 | 467 | 61 | 2 | 252 | 127 | .. | 434 | 224 | 20 | 86 |
| R.A.F. 3a | 6 | 117 | .. | 2 | 1 | .. | 13 | 1 | .. | 5 |
| 70/80-h.p. Renault | 2 | 688 | 741 | 196 | 12 | .. | 373 | 29 | .. | .. |
| 160/190-h.p. Renault | .. | .. | 17 | 8 | .. | .. | 12 | 10 | .. | .. |
| 220/240-h.p. Renault | 28 | 4 | 64 | .. | .. | .. | 9 | 3 | .. | .. |
| Rolls-Royce 'Falcon' | 289 | 39 | .. | 91 | 46 | 3 | 138 | 91 | 69 | .. |
| Rolls-Royce 'Eagle' (Series 1–7) | 26 | 105 | .. | 4 | 26 | .. | 44 | 22 | .. | .. |
| Rolls-Royce 'Eagle' (Series 8) | 76 | 45 | .. | 283 | 147 | 7 | 346 | 65 | 1 | .. |
| 135/150-h.p. Salmson | .. | .. | 156 | .. | .. | .. | .. | .. | .. | .. |
| 200-h.p. Salmson | .. | .. | 17 | .. | .. | .. | .. | .. | .. | .. |
| Siddeley 'Puma' | 519 | 176 | .. | 561 | 172 | 54 | 560 | 276 | 1 | 56 |
| 150/170-h.p. Sunbeam | .. | .. | 222 | .. | .. | .. | .. | 3 | .. | .. |
| Sunbeam 'Arab' | 218 | 9 | .. | 270 | 97 | .. | 96 | 71 | .. | .. |
| 225/240-h.p. Sunbeam | .. | 1 | 223 | .. | .. | .. | 2 | .. | .. | .. |
| Sunbeam 'Maori' | 115 | 64 | .. | 137 | 69 | .. | 206 | 59 | .. | .. |
| Sunbeam 'Cossack' | 86 | 12 | 64 | 1 | 2 | .. | 10 | 3 | .. | .. |
| Wolseley 'Adder' | 66 | 7 | .. | 2 | 4 | .. | 33 | 15 | 7 | .. |
| Wolseley 'Python' | 38 | 25 | .. | 2 | 1 | .. | 55 | 20 | .. | .. |
| Wolseley 'Viper' | 201 | 53 | .. | 132 | 104 | 9 | 44 | 25 | 1 | .. |
| Miscellaneous | .. | 8 | 67 | .. | .. | .. | 3 | .. | 1 | .. |
| | 5,090 | 2,139 | 2,741 | 3,942 | 1,553 | 111 | 7,455 | 2,713 | 936 | 352 |

\* Includes 1109 engines in workshops for repair.
† Includes all engines not advised as received by Mediterranean and M.E.B. Stations.

# ENGINES

| C.T.D. | Grand Fleet and Northern Patrol. | Expeditionary Forces. | | | | Eastern Stations. | | | | | Mediterranean (30th Sept. 1918). | Total. | Engines written off charge during October. |
| | | E.F. France. | I.F. France. | 14th Wing Italy. | 5th Group. | Awaiting Shipment and in Transit to the East. | Egypt and Palestine. | Salonika. | Mesopotamia. | N.W. Frontier. | | | |
|---|---|---|---|---|---|---|---|---|---|---|---|---|---|
| 18 | .. | .. | .. | .. | .. | .. | .. | .. | .. | .. | .. | 19 | .. |
| 13 | .. | .. | .. | .. | .. | .. | .. | .. | .. | .. | .. | 13 | .. |
| .. | .. | .. | .. | .. | .. | .. | .. | .. | .. | .. | .. | 241 | .. |
| 2 | .. | .. | .. | .. | .. | .. | 16 | .. | 1 | 1 | .. | 162 | 1 |
| 17 | .. | 617 | .. | .. | .. | 29 | 108 | 76 | 10 | 3 | .. | 1,754 | 43 |
| 6 | 171 | 340 | .. | .. | 22 | .. | .. | .. | .. | .. | 7 | 833 | 28 |
| 30 | 6 | 149 | 3 | .. | .. | .. | .. | .. | .. | .. | .. | 718 | 2 |
| 4 | .. | .. | .. | .. | .. | 17 | 113 | .. | .. | .. | .. | 322 | 3 |
| 30 | 113 | 11 | 4 | 104 | 4 | .. | 29 | .. | .. | .. | 152 | 2,951 | 6 |
| 12 | 21 | 466 | 23 | .. | 9 | 31 | .. | .. | .. | .. | 43 | 914 | 35 |
| 2 | .. | .. | .. | .. | .. | .. | .. | .. | .. | .. | .. | 4 | .. |
| 9 | .. | .. | .. | .. | .. | .. | .. | .. | .. | .. | .. | 659 | 5 |
| .. | .. | .. | .. | .. | .. | .. | .. | .. | .. | .. | .. | 46 | .. |
| 9 | .. | 2 | .. | .. | .. | .. | .. | .. | .. | .. | .. | 139 | 1 |
| 2 | .. | .. | .. | .. | .. | .. | .. | .. | .. | .. | .. | 89 | .. |
| 5 | .. | .. | .. | .. | .. | .. | .. | .. | .. | .. | .. | 25 | .. |
| 9 | .. | .. | .. | .. | .. | .. | 136 | .. | .. | .. | .. | 1,000 | 5 |
| 39 | .. | 532 | .. | .. | .. | 16 | 36 | 1 | .. | .. | .. | 2,173 | 35 |
| .5 | .. | 2 | .. | .. | .. | .. | .. | .. | .. | .. | .. | 21 | .. |
| 24 | 91 | 2 | .. | 1 | .. | .. | .. | .. | .. | .. | 12 | 1,708 | 5 |
| 13 | 25 | 547 | .. | .. | .. | 26 | 116 | 42 | 42 | .. | .. | 2,589 | 22 |
| 14 | 14 | 77 | 78 | .. | .. | .. | .. | .. | .. | .. | .. | 622 | 11 |
| 5 | 2 | .. | .. | .. | .. | 90 | 195 | .. | .. | .. | 2 | 1,854 | 2 |
| 7 | .. | .. | .. | .. | .. | .. | .. | .. | .. | .. | .. | 22 | .. |
| 25 | 12 | 8 | .. | .. | .. | .. | 282 | 21 | 10 | 75 | 10 | 2,253 | 48 |
| 23 | 1 | 818 | .. | 61 | .. | 34 | 265 | 12 | 78 | .. | .. | 2,996 | 33 |
| 4 | .. | 42 | .. | .. | .. | .. | .. | .. | 22 | .. | .. | 229 | 1 |
| 9 | .. | .. | .. | .. | .. | .. | 61 | .. | .. | .. | .. | 2,117 | .. |
| .. | .. | .. | .. | .. | .. | .. | .. | .. | .. | .. | 11 | 47 | 13 |
| .. | 7 | .. | .. | .. | .. | .. | .. | .. | .. | .. | .. | 126 | .. |
| 7 | 4 | 250 | .. | 57 | .. | 27 | 70 | .. | .. | .. | .. | 1,182 | 20 |
| 6 | 26 | 170 | .. | .. | 10 | .. | .. | .. | .. | .. | 3 | 451 | 8 |
| 32 | 45 | 187 | 284 | .. | 30 | 32 | 6 | .. | .. | .. | 34 | 1,626 | 28 |
| 3 | .. | .. | .. | .. | .. | .. | 20 | .. | .. | .. | 8 | 187 | 13 |
| 8 | .. | .. | .. | .. | .. | .. | .. | .. | .. | .. | 5 | 30 | .. |
| 40 | 3 | 462 | 76 | .. | 8 | 120 | 38 | 27 | .. | .. | 102 | 3,255 | 63 |
| 1 | .. | .. | .. | .. | .. | .. | 1 | .. | .. | .. | 6 | 233 | .. |
| 25 | 53 | 126 | .. | .. | .. | .. | .. | .. | .. | .. | .. | 967 | 2 |
| 3 | .. | .. | .. | .. | .. | .. | .. | .. | .. | .. | 9 | 238 | .. |
| 9 | 14 | .. | .. | .. | 10 | 24 | 11 | .. | .. | .. | 82 | 800 | 3 |
| 4 | .. | .. | .. | .. | .. | 2 | .. | .. | .. | .. | 51 | 235 | 3 |
| 5 | .. | 48 | .. | .. | .. | 34 | 63 | 23 | 21 | .. | .. | 328 | 2 |
| 6 | .. | 1 | .. | .. | .. | .. | .. | .. | 13 | .. | .. | 161 | .. |
| 10 | .. | 683 | .. | .. | .. | .. | .. | .. | .. | .. | .. | 1,263 | 58 |
| 17 | .. | .. | .. | .. | .. | .. | .. | .. | .. | .. | .. | 100 | 50 |
| 522 | 608 | 5,540* | 468 | 223 | 93 | 482† | 1,566 | 202 | 197 | 79 | 537 | 37,702 | 549 |

www.ingramcontent.com/pod-product-compliance
Lightning Source LLC
Chambersburg PA
CBHW030403100426
42812CB00028B/2822/J